WHAT EVERY ENGINEER SHOULD KNOW ABOUT

BUSINESS
COMMUNICATION

WHAT EVERY ENGINEER SHOULD KNOW
A Series

Series Editor*
Phillip A. Laplante
Pennsylvania State University

*Founding Series Editor: **William H. Middendorf**

WHAT EVERY ENGINEER SHOULD KNOW ABOUT

BUSINESS COMMUNICATION

John X. Wang

CRC Press
Taylor & Francis Group
Boca Raton   London   New York

CRC Press is an imprint of the
Taylor & Francis Group, an **Informa** business

CRC Press
Taylor & Francis Group
6000 Broken Sound Parkway NW, Suite 300
Boca Raton, FL 33487-2742

© 2008 by Taylor & Francis Group, LLC
CRC Press is an imprint of Taylor & Francis Group, an Informa business

No claim to original U.S. Government works

International Standard Book Number-13: 978-0-8493-8396-0 (Softcover)

Library of Congress Cataloging-in-Publication Data

Wang, John X., 1962-
    What every engineer should know about business communication / John X. Wang.
        p. cm. -- (What every engineer should know)
    Includes bibliographical references and index.
    ISBN-13: 978-0-8493-8396-0 (alk. paper)
    ISBN-10: 0-8493-8396-X (alk. paper)
    1. Communication in engineering. 2. Business communication. 3. English
language--Business English. I. Title.

TA158.5.W35 2008
658.4'5--dc22   2007049508

Visit the Taylor & Francis Web site at
http://www.taylorandfrancis.com

and the CRC Press Web site at
http://www.crcpress.com

To the Mississippi

I crossed the river hundreds of times when writing this new book.

Good business communication flows just like the great river.

Contents

Section 2: Write Your Way for Business Impact

Section 3: Integrating Your Speaking and Writing Skills

Preface

Engineers require an ever-increasing range of business communication skills to maintain relevance with the dynamic business environment. Engineers in all positions must communicate the purpose and relevance of their work, both orally and in writing.

If you work in industry, you must communicate with managers and co-workers, and perhaps with customers and suppliers. If you are responsible for raising funds for your Research & Development (R&D) project, you must market your ideas effectively, write proposals, and generate enthusiasm for your research. If you work in public policy or government, you might have to communicate with the press and other members of the public. Employers now seek graduates with excellent skills that go beyond academic credentials, including:

- Business communication
- Decision making in uncertainty
- Risk engineering and management

Engineers use business communication skills to explain an idea, process, or technical design. Volumes 36 and 37 of the *What Every Engineer Should Know* series dealt with decision making in uncertainty, and risk engineering and management. Although many executives are pleased with the basic technical skills their new engineers bring to the organization, they are concerned that many engineers lack the critical business communication skills necessary in today's workplace. Today's engineering executives want engineers who can communicate clearly, concisely, and comprehensively. This new volume of the *What Every Engineer Should Know* series discusses how engineers can use written and oral skills, computers, graphics, and other engineering tools to communicate with other engineers and management.

The knowledge of business communication skills is quite different from the knowledge of communication theory. Because the knowledge of communication theory does not necessarily parallel skills in practice, it is important to immerse engineers in similar work environments. Context-specific enactments, or role-play, can focus the engineer's attention on the differing types of communication required with various groups in potential work situations. By engaging the engineers directly in active learning, the book is organized as follows:

Part 1—Speak Your Way to Engineering Success: Oral communication helps you to deliver presentations, explains a design or design process, improves meeting coordination, and develops a project team.

Part 2—Write Your Way to Business Impact: Written communication helps you write technical reports, specifications, and other informational material. It includes specific, to-the-point details about a topic that you want to communicate. Other engineers use these documents for continued research and development because your knowledge has been communicated to others.

Part 3—Integrating Your Speaking and Writing Skills: This requires integrating speaking skills, writing skills, visualization skills, and listening stills into everyday communication, technical proposals and reports, and risk communication. Visualization encompasses a large range of topics from rough, preliminary sketches and visual aids to detailed computer CAD diagrams. Marketing-related knowledge for proposal development is also crucial to all stages of the design process.

Business communication skills basically constitute several core elements, such as the fluency in the oral/written communication and the fundamentals of visual communication. These skills are essential for an engineer who aspires to pursue his or her professional career in the global arena of a dynamic business environment. I sincerely hope this book can help engineers acquire the essential skills for professional success.

John X. Wang, Ph.D.
Marion, Iowa

About the Author

John X. Wang, Ph.D., is the founder and Chief Master Black Belt of Lean Six Sigma Institute of Technology, Marion, Iowa. He has taught engineering training courses at Panduit Corporation, Maytag Corporation, and Visteon Corporation. Dr. Wang has taught reliability engineering and design for Six Sigma Manager at Maytag Corporation (where he led reliability engineering best practices and Design for Lean Six Sigma training), as a Six Sigma Master Black Belt certified by Visteon (where he led Design for Six Sigma training programs), and as a Six Sigma Black Belt certified by General Electric (where he led Design for Six Sigma best practice projects).

Dr. Wang has authored and coauthored three engineering books and numerous professional publications on decision making under uncertainty, risk engineering and management, Six Sigma, reliability engineering, and systems engineering. Dr. Wang has also taught engineering and professional courses at Gannon University, Erie, Pennsylvania, and National Technological University, College Park, Maryland. He has spoken and presented at various international and national engineering conferences, symposiums, professional meetings, seminars, and workshops.

Dr. Wang has been designated as a Certified Reliability Engineer by the American Society for Quality and as a Certified Master Black Belt by the International Quality Federation. He received a B.A. in engineering physics in 1985 and an M.S. in system engineering and physics from Tsinghua University, Beijing, China. In 1995, Dr. Wang received his Ph.D. from the University of Maryland at College Park.

Dr. Wang lives in Marion, Iowa, with his wife and two sons.

1

Analyze Communication Purpose and Audience

1.1 How Engineers Learn

The art of getting your message across is a vital part of being a successful engineer. Whether you want to make a presentation to the design review meeting or to negotiate the product specifications with suppliers, you need to understand how engineers learn.

Engineers seek optimal solutions to problems. Often, however, the constraints of the problem and the solution criteria are of several, qualitatively different types, and there is no formal way to find the best trade offs. Nevertheless, engineers make judgments and provide explanations to justify their choices. Engineering communication is the development of such explanations that identify and validate a particular solution as the best. Engineers' thinking involves analogical reasoning as well as deduction. This implies that in engineering communication, descriptive case-based examples are important as source analogs for problem solving.

Engineers often learn from colleagues through communication. For engineers, communication is an important way of learning, which can be defined formally as the act, process, or experience of gaining knowledge or skills. Communication and the subsequent learning help engineers move from novices to experts and allow them to gain new professional knowledge and abilities.

In today's business environment, finding better ways to communicate will propel organizations forward. Strong communication links fuel strong organizations. To accomplish effective business communication, we must capitalize on natural learning styles preferred by engineers. As presented in Figure 1.1, a better understanding of business communication should lead not only to a better understanding of ourselves as engineers, but also lead to improved working relationships and team dynamics.

Good communication is the lifeblood of the engineering profession. It takes many forms, such as speaking, writing, and listening. Use it to handle information, persuade people, and improve relationships.

Business communication has a profound effect upon working relationships and job effectiveness. An engineer's communication style includes far more than the words and sentences. Each communication is composed of contextual cues, unspoken messages, and the rhythms of speaking or writing.

FIGURE 1.1
Business communication shapes our lives and careers in numerous ways.

1.1.1 Recognize That Every Engineer Is Unique

Effective engineering communication hinges on engineers understanding the meaning of your message. They learn best when their uniqueness in the engineering profession is considered. Business communication will be ineffective if the uniqueness is neglected. Further, an engineer learns best when her or his beliefs, emotions, values, and needs are considered. The more we demonstrate awareness of an engineer's needs, wants, and desires, the more successful the business communication will be. Engineers are *goal-oriented*. Therefore, they appreciate a business communication that is organized and has clearly defined elements. You, as the presenter, must show engineers how the communication will help them attain their goals. This classification of goals and objectives must be done early in the business communication.

1.1.2 Gain Trust

Engineers learn best when they can concentrate on learning and communicating instead of defending against rejection, anxiety, shame, fear of failure, or uncertain expectations. Engineers learn best in situations that reinforce self-esteem and maximize possibilities for success. Good communication means saying what you mean—and fully comprehending any feedback. The following three rules will help you to gain engineers' trust:

1. Be clear in your own mind about what you want to communicate.
2. Deliver the message succinctly.
3. Ensure that the message has been clearly and correctly understood.

1.1.3 Solve Engineering Problems

Communication involves at least two parties. Each of them may have different wants, needs, desires, and attitudes. These wants, needs, and desires

What is the engineering problem?

How frequently is the problem occurring?

What is the effect of the problem?

How severe is the effect?

Why is the problem happening?

How to prevent the problem?

Where to validate your solution?

FIGURE 1.2
Solve engineering problems.

can present barriers if they conflict with those of the other party. You must overcome these barriers to communicate effectively. As described in Figure 1.2, ask a series of questions to discover the meaning of a particular problem or situation.

Engineers are *relevancy-oriented*. An engineer learns best if she or he is highly motivated to learn. She or he also needs a reason to acquire new knowledge or change an existing view about an engineering product, process, or service. An engineer's motivation to learn is high when an unmet need for product/process/service improvement is determined. Therefore, we must identify objectives for engineers before the communication begins. In addition this means that theories and concepts must be related to a setting familiar to engineers. Encouraging engineers to choose projects that reflect their own interests often fulfills this well.

1.1.4 Respect Experience and Ability

An engineer learns best when her or his previous learning and professional experiences are considered. Whenever possible, business communication should build on previous learning and experience. Repetition during business presentation or training should refresh or reinforce prior experience. During the communication process, we must acknowledge the wealth of experiences that engineers bring to the table. These engineers should be treated as equals in experience and knowledge and allowed to voice their opinions freely during the business communication process.

Asking open-ended questions and listening attentively to the answers is a great way to show your respect for their experience and ability. Open-ended questions should begin with words such as "why" and "how" or phrases such as "What do you think about … ?" Open-ended questions should lead engineers to think analytically and critically. Open-ended questions encourage engineers to think, to express opinions, and to share ideas. They indicate your wish for such an open response. Ultimately, a good open-ended question should stir discussion and spark enthusiasm and energy in the engineers.

1.1.5   Control the Learning Experience

Engineers should be involved in assessing needs, and planning, delivering, and evaluating their learning experiences from business communications, when possible. Engineers are often autonomous and self-directed. They need to be free to direct themselves. An effective business communication often involves the engineering audience to assume responsibility for presentations and group leadership. These calls for learning are:

- Hands-On—Engineers are actually allowed to perform hands-on practices as they construct meaning and acquire understanding.
- Minds-On—Activities focus on core concepts, allowing engineering to develop thinking processes and encouraging them to question and seek answers that enhance their knowledge.
- Authentic—Engineers are presented with problem-solving activities that incorporate authentic, real-life questions and issues in a format that encourages collaborative effort, dialogue with informed expert sources, and generalization to broader ideas and application.

1.1.6   Allow Time to Alter Perceptions

Engineers have accumulated a foundation of life experiences and knowledge that may include work-related activities, family responsibilities, and previous education. They need to connect learning to this knowledge/experience base. Engineers bring feelings about previous learning and professional experience, attitudes, beliefs, and at least some resistance to change with them. To help them appreciate the value of the new information, we should draw out engineers' experiences and knowledge that are relevant to the topic.

Empathy helps accomplish effective communication with engineers. Empathy is commonly defined as one's ability to recognize, perceive, and directly experientially feel the emotion of another. As the states of mind, beliefs, and desires of others are intertwined with their emotions, one with empathy for another may often be able to more effectively divine another's modes of thought and mood. Empathy is often characterized as the ability to "put oneself into another's shoes," or to experience the outlook or emotions of another being within oneself—a sort of emotional resonance.

1.1.7   Hold the Engineer's Interest

An engineer learns best when she or he is involved in the process. The quickest way to lose an engineer's attention is to lecture or use other passive (noninteractive) delivery methods during business communication.

Remember, we must relate theories and concepts to the engineers and recognize the value of their experience in learning. To hold engineers' interest, you need to focus on their interests for engineering problem solving.

Engineering problems come in all sizes, shapes, and colors. There is no single or simple step-by-step process that guarantees we will solve every problem encountered. We are faced instead with the requirement to configure or adapt our problem-solving processes to fit the problem at hand. Help engineers to enter the creative problem-solving cycle with a statement of a problem that pertains to a specific topic. It may take any of these forms:

- An open-ended question
- An existing condition in need of change
- A kit of materials provided to design and build something
- A research proposal for a student project

  Example: A self-described "ecologically concerned" client wants to change her home heating system from electricity to gas to dramatically reduce her heating costs.

1.1.8 Present Meaningful Contents

Engineers are practical, focusing on the aspects of communication most useful to them in their work. They may not be interested in knowledge for its own sake. We must tell engineers explicitly how the communication will be useful to them on the job. An engineer learns best if content is meaningfully communicated. To involve engineers, use real engineering examples and case studies.

Case study helps capture the engineers' attention. Engineers often think with case-based reasoning. To solve an engineering problem, the problem is matched against the cases in the case base, and similar cases are retrieved. The retrieved cases are used to suggest a solution that is reused and tested for success. If necessary, the solution is then revised. Finally, the current problem and the final solution are retained as part of a new case. Engineers prefer case study and case-based reasoning because they can relate to specific examples instead of conclusions that may not relate specifically to their own experiences. A case-study library can also be a powerful corporate resource, allowing everyone in an organization to tap into the corporate case-study library when handling a new problem.

1.2 How Engineers Are Persuaded

Persuasion is a form of influence. It is the process of guiding your project team toward the adoption of an idea, attitude, or action by effective business

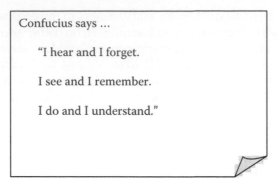

Confucius says …

"I hear and I forget.

I see and I remember.

I do and I understand."

FIGURE 1.3
"I do and I understand!"

communication. It is a problem-solving strategy for engineering projects. How does one persuade other engineers? As Confucius told us, communication through hands-on experience is very powerful (see Figure 1.3). This is true for both ancient engineers and modern engineers. In general, we remember

- 10% of what we read
- 20% of what we hear
- 30% of what we see
- 50% of what we see *and* hear
- 70% of what we say or write (speaking and writing are good learning methods)
- 90% of what we say and do (case studies with discussion, hands-on activities, engineering simulations, and role-plays)

Effective business communications require the following:

- Presentations that contain many examples or stories that demonstrate application of material
- Appropriate demonstrations, experiments, practical problems, and case studies to facilitate communications
- Interaction through questions and answers
- Focus on the most important communication objectives, that is, prioritize communication contents based on the "80/20 rule."

When applicable, simulation-based, hands-on exercises help to enhance engineers' learning experiences. A simulation is an imitation of real things, states of affairs, or processes. The act of simulating something generally entails representing certain key characteristics or behaviors of a selected engineering system. Simulation is used in many contexts, including the

> When the facts support your design,
>
> present the facts.
>
> When the theory supports your
>
> design, present the theory.
>
> And when neither the facts nor the
>
> theory support your design, pound on
>
> the drawing board!

FIGURE 1.4
An engineering poem illustrating Aristotle's logos, ethos, and pathos.

modeling of natural systems or human systems to gain insight into their functioning. Other contexts include simulation of technology for performance optimization, safety engineering, testing, training, and education. Simulation can be used to show the eventual real effects of alternative conditions and courses of action.

All business communication contains some elements of logic, ethics, and emotion. Aristotle recognized this in his *The Art of Rhetoric*, noting that all communication contains "logos," "ethos," and "pathos" (see Figure 1.4). Created by Edward de Bono, "Six Thinking Hats" is an important and powerful technique for business and engineering thinking. It can be used to look at business communication from a number of important perspectives. This forces you to move outside your habitual thinking style and helps you to get a more rounded view about business communication. You can use Six Thinking Hats in presentations, meetings, reports, or other business communications. It has the benefit of blocking the confrontations that happen when people with different thinking styles discuss the same problem. Each "Thinking Hat" is a different style of thinking as explained next:

1. White Hat—This covers facts, figures, information needs, and gaps. With this thinking hat, you focus on the data available. Look at the information you have, and see what you can learn from it. Look for gaps in your knowledge, and either try to fill them or take account of them. This is where you analyze past trends and try to extrapolate from historical data.

2. Red Hat—This covers intuition, feelings, and emotions. "Wearing" the Red Hat, you look at problems using intuition, gut reaction, and emotion. In addition, try to think how other people will react

emotionally. Try to understand the responses of people who do not fully know your reasoning. The red hat allows the thinker to put forward an intuition without any need to justify it.

3. Black Hat—This is the hat of judgment and caution. Using Black Hat thinking, look at all the bad points of the project, product, or process. Look at it cautiously and defensively. Try to see why it might not work. This is important because it highlights the weak points in a plan. It allows you to eliminate them, alter them, or prepare contingency plans to counter them. Black Hat thinking helps to make your plans "tougher" and more resilient. It can also help you to spot fatal flaws and risks before you embark on a course of action. Black Hat thinking is one of the real benefits of this technique, because many successful engineers get so used to thinking positively that often they cannot see problems in advance. This leaves them underprepared for difficulties.

4. Yellow Hat—This is the logical positive. The Yellow Hat helps you to think positively. It is the optimistic viewpoint that helps you to see all the benefits of the project, product, or process. Yellow Hat thinking helps you to keep going when everything looks gloomy and difficult. It can be used in looking forward to the results of some proposed action, but can also be used to find something of value in what has already happened.

5. Green Hat—The Green Hat stands for creativity. This is the hat of creativity, alternatives, proposals, interesting aspects, provocations, and changes. This is where you can develop creative solutions to a problem. It is a freewheeling way of thinking in which there is little criticism of ideas. A whole range of creativity tools can help you here.

6. Blue Hat—The Blue Hat stands for process control. Used to manage the thinking process, this is the hat worn by people chairing meetings. So, when running into difficulties because ideas are running dry, the chairperson may direct activity into Green Hat thinking, or when contingency plans are needed, he or she will ask for Black Hat thinking, and so forth.

The Six Thinking Hats technique is useful for looking at the effects of a communication approach from a number of different points of view. It allows necessary emotion and skepticism to be brought into what would otherwise be purely rational decisions. It opens up the opportunity for creativity within business communication. The technique also helps, for example, persistently pessimistic people to be positive and creative. Business communication developed using the Six Thinking Hats technique will be sounder and more resilient than would otherwise be the case. It may also help you to avoid miscommunications. The method promotes fuller input from more people. In de Bono's words, it "separates ego from performance." Everyone is able to contribute to the exploration without denting egos as they are just using the Yellow Hat or whatever color hat. The six hats system encourages performance instead of ego defense. People can contribute under any hat, although they might initially support the opposite view.

1.3 Speak or Write: Select the Right Communication Channel

To communicate effectively, it is essential that you choose the suitable medium. Selecting the right communication channel is like shopping in the right store. If you select the wrong store, you will not get the items you want. Similarly, if you select the wrong communication channel, you will not get the results you want.

For many engineers, the choice is often between the spoken and the written word. Do you have to write? It is hard work, time-consuming, and expensive. A phone call or face-to-face meeting may be better, especially if give and take is needed or the subject matter is sensitive. If you decide that you want speed and convenience, you may well choose speech as the best form of communication. For a record of the conversation, you can follow with a confirming e-mail or meeting minute.

Would speaking or writing be appropriate? Sometimes you just have to write. You may want something more permanent and orderly—a typed request, for example—which will elicit a specific reply. Your ideas may be too complex to handle orally. You may have to reach too many people in too many locations. The managers may insist on documentation for future reference. The nature of the engineering task may call for a proposal, plan, report, or specification in writing. Finally, your physical or psychological distance from the audience may leave you no choice but to write.

When you must write about routine and repetitive subjects, consider using form letters or paragraphs stored in word processing files. They affect many readers, so craft them initially and update them regularly. Also remember that a brief e-mail or note can handle many routine matters simply and cheaply. E-mails have the speed and informality of a phone conversation, yet they can be filed. The purpose of your message will dictate which method to choose (see Figure 1.5).

1.4 Consider Your Communication Purpose and Audience

Before presenting a speech or a written communication, analyze your audience or readers in terms of your purpose.

Writing is hard work. It is time

consuming and expensive.

A conversation may be better,

especially if negotiation is needed or

the subject is sensitive.

FIGURE 1.5
Do you have to write?

- What is your main purpose? You should be able to state it in a sentence or two. Use the main purpose to guide everything you say or write. Should you inform or persuade? Informative communication gives straight information to whoever is listening or reading. For persuasive communication, to communicate something is to do something.

- Who is the key audience or reader? Many people may review your speech or written communication before it is presented or distributed. If you try to satisfy everyone, you will satisfy no one.

- How will they use what you say or write? The answer affects everything: the content, organization, and wording.

- What is your relationship to your audience or readers? The closer that relationship, the more you can relax your tone and expect some benefit of the doubt if you do not explain yourself as fully as you might want to.

- How much do your audience or readers know already? You will bore them by dwelling on what they know already and confuse them by assuming too much.

- What do they need to know now? Anticipate the likely questions, and answer them in advance.

- What will make it easy for them to understand or act? To be easy on your audience or readers, you must be hard on yourself.

- How unusual is your information? The more unusual, the more you will have to explain, especially if you are presenting designs and manufacturing processes that are brand-new. Claude Shannon, an American electrical engineer and mathematician, proved the relationship between "low probability of surprise" and "large amount of information" in his information theory.

Pay careful attention to these questions, no matter how basic they may seem. Ignoring them would cause trouble, even before the first word is spoken or written. The answers call for imagination: Step back from your own interests, and see things from other angles. To speak or write well, you do not need to agree with your audience and readers. What is essential is to see the world from their perspectives.

Bibliography

Cartwright, R. and NetLibrary, Inc., *Communication*. Oxford, UK: Capstone Publishing, 2002.

Chaney, L. H. and Martin, J. S., *Intercultural Business Communciation*. 2nd ed. Upper Saddle River, NJ: Prentice Hall, 2000.

Cialdini, R. B., *Influence: The Psychology of Persuasion*. Rev. ed. New York: Morrow, 1993.

Downey, R., Boland, S., and Walsh, P., *Communications Technology Guide for Business*. Boston: Artech House, 1998.

Eckhouse, B. E., *Competitive Communication: A Rhetoric for Modern Business*. Rev. ed. New York: Oxford University Press, 1999.

Fearn-Banks, K., *Crisis Communications: A Casebook Approach*. 2nd ed. Mahwah, NJ: Lawrence Erlbaum Associates, 2002.

Gardner, H., *Changing Minds: The Art and Science of Changing Our Own and Other People's Minds*. Boston: Harvard Business School Press, 2004.

Harvard Business Essentials: Business Communication. Boston: Harvard Business School Press, 2003. (The Harvard Business Essentials series.)

Harvard Business Review on Effective Communication. Boston: Harvard Business School Press, 1999. (The Harvard Business Review paperback series.)

Holtz, S., *Corporate Conversations: A Guide to Crafting Effective and Appropriate Internal Communications*. New York: AMACOM, 2004.

Isaacs, W., *Dialogue and the Art of Thinking Together: A Pioneering Approach to Communicating in Business and in Life*. New York: Currency, 1999.

Jablin, F. M. and Putnam, L., *The New Handbook of Organizational Communication: Advances in Theory, Research, and Methods*. Thousand Oaks, CA: Sage Publications, 2001.

Kenton, S. B and Valentine, D., *Crosstalk: Communicating in a Multicultural Workplace*. Upper Saddle River, NJ: Prentice Hall, 1997.

Levine, R. V., *The Power of Persuasion: How We're Bought and Sold*. Hoboken, NJ: John Wiley & Sons, 2003.

Munter, M., *Guide to Managerial Communication: Effective Business Writing and Speaking*. 4th ed. Upper Saddle River, NJ: Prentice Hall, 1997.

Nolan, R. W., *Communicating and Adapting across Cultures: Living and Working in the Global Village*. Westport, CT: Bergin & Garvey, 1999.

Pan, Y., Scollon, S. B. K., and Scollon, R., *Professional Communication in International Settings*. Malden, MA: Blackwell Publishers, 2002.

Pearce, T., *Leading out Loud: Inspiring Change through Authentic Communication*. New and rev. ed. San Francisco: Jossey-Bass Publishers, 2003.

Rosner, B., Halcrow, A., Levins, A. S., and NetLibrary, Inc., *Communication*. New York: McGraw-Hill, 2001.

Simmons, J., *We, Me, Them, & It: The Powers of Words in Business*. New York and London: Texere, 2002.

Whalen, D. J., *I See What You Mean: Persuasive Business Communication*. Thousand Oaks, CA: Sage Publications, 1996.

Wiener, V., *Power Communications: Positioning Yourself for High Visibility*. New York: New York University Press, 1994.

1

Speak Your Way to Engineering Success

2

Projecting the Image of the Engineering Profession

As an engineer, you want to be described as technically competent, socially skilled, of strong character and integrity, and committed to your work, your team, and your company. Research shows that the most favorably regarded traits are trustworthiness, caring, humility, and capability. Your professional image affects your reputation, and your reputation affects your success. The image you project during business communication will directly influence your performance. Although your working style will always retain your personal touch, you can observe a few general guidelines and suggestions to help project a positive and respectable image to your customers, management, engineering associates, and suppliers.

Projection is nine tenths of success. Studies indicate that communication skills impact engineers' effectiveness and success more than any other skill, including technical knowledge. Furthermore, all the studies indicate that the impression you make as an engineering professional is far more important than the content of what you actually say. So, why do we spend so much time working on the content of our presentation with little attention to our delivery styles, techniques, and environment? Yes, the content must be sound, accurate, well prepared, and suitable to the level of the audience; delivery techniques are equally important. Superior delivery techniques can facilitate a challenging engineering presentation, communicating the technical contents smoothly.

Projecting a professional and positive image is a product of good presentation etiquette, strong physical and vocal skills, and quality content. How you look also communicates a lot about you. Compliment your audience by what you wear. You have probably dressed appropriately if you overhear some saying, "That must be the speaker." Personal grooming is even more important than what you wear. There is no right way to dress—only an appropriate way. With regard to presentation, it is imperative to pay attention to details and to always be enthusiastic about your topic.

2.1 Overcome Anxiety

Speech anxiety can have a negative effect on careers and the ability to get things done. It may be a lifelong fear or current apprehension toward a specific situation. Either way, it can severely limit personal and organizational goals, including career advancement, company outreach efforts, and visibility for you and your organization.

Anxiety is a natural state that exists any time we are placed under stress. If you are nervous, you are in good company. Among the most stressful situations people encounter, speaking before a group tops the list. When this type of stress occurs, fears take place that may cause symptoms such as a nervous stomach, sweating, tremors in the hands and legs, accelerated breathing, or increased heart rate.

2.1.1   Why Are We Afraid of Making Presentations?

The reason most people get anxious when required to speak to a group is that they are afraid of looking foolish or stupid in front of many of their peers and important people. They are afraid that their minds will go blank or that their lack of speaking skills will lower the opinion others have of them.

2.1.2   Steps You Can Take to Reduce the Fear

Don't worry! Anxiety is normal. Almost everyone experiences some stress before speaking. Coaches of sports teams *want* their players to be anxious before a game. Anxiety produces energy and excitement. In most cases, the fear or nervousness is just extra energy. This "extra energy" can be incorporated in the speech if it is controlled; however, you first must attempt to reduce the fear. Anxiety that produces adrenaline and enhances your presentation is desirable, and anxiety that hinders your performance must be managed. The trick is to make your excess energy work for you. The easiest way to do this is through preparation.

As described next, you can use several steps and tricks to reduce the fear of making a mistake or looking foolish when you speak to a group.

2.1.2.1   *Prepare, Prepare, and Prepare*

Of course, you are in for a long, uphill battle if you are unprepared, disorganized, or late. Lack of organization is one of the major causes of anxiety. These are all problems that can be eliminated with careful planning. Knowing that your thoughts are well organized will give you more confidence, which will allow you to focus energy into your presentation.

One of the best ways to make sure you do not make foolish mistakes is to be well prepared before you speak to a group. This does not mean that you should memorize exactly what you plan to say. Instead, you should have a good outline of facts and information ready for your presentation.

A professional in any field does not leave anything to chance before a big game, an important performance, or a critical presentation to corporate executives. Strategies are laid out, all material is ready, contingency plans are made, and every detail is considered. When we do not prepare, we prepare to fail. When you are well prepared, chances of failure or mistakes are greatly reduced. You feel more relaxed and sure of yourself, because you have all the bases covered.

2.1.2.2   Have a Backup

It is worthwhile to bring along a "security blanket" or "safety net" in case something goes wrong with your presentation. For example, having your speech outlined on some sort of cards or pages is a good backup in case you have a mental lapse. Referring to your notes is certainly acceptable to refresh your memory; however, you should be prepared enough that you do not have to completely depend on your notes for your material. Do not use your notes/speaking outline as a crutch, but to keep you on track.

2.1.2.3   Reduce Your Fear of the Audience

Speaking to peers, employers, instructors, or dignitaries can create fear in a person. Try alleviating that fear by greeting your audience at the door. Getting to know them early will help you realize that they are also engineering professionals just like you. Think about your audience as being on "your team." Become one with your audience. Do not build an artificial wall between you and your audience. Direct eye contact can create a oneness between you and your audience. Use the abundance of energy that your audience is capable of projecting to you. To paraphrase the recurring theme from the movie *Star Wars:* Let the audience be with you. Remember that they are on your team, and they are there to hear what you have to say. Speakers are often nervous or frightened of the unknown.

2.1.2.4   Practice, Practice, and Practice

Even if you know your material very well, practice is extremely important. The more you give a talk, the more automatic it becomes, the more energy it can have, and the more confidence you have in your abilities to give the speech. Practice out loud, alone, to small groups, to friends, to relatives, to strangers, to pets, and to roommates. Treat your practice just like you treat the speech on speech day. For example, if you want to have energy and enthusiasm on speech day, then you will want to practice with energy and enthusiasm.

Some other techniques for managing fear include stretching exercises, deep breathing (see Figure 2.1), brisk walks, and avoiding coffee and alcohol. When your muscles tighten and you feel nervous, you may not be breathing deeply enough. The first thing to do is to sit up, erect but relaxed, and inhale deeply a number of times. Instead of thinking about the tension, focus on being confident. As you breathe, tell yourself on the inhale, "I am" and on the exhale, "confident." Try to clear your mind of everything except the repetition of the "I am … confident" statement, and continue this for several

- Sit down or lie down.

- Inhale slowly and say to yourself

  I am...

- Exhale slowly and say to yourself

  confident.

FIGURE 2.1
Deep Breathing: a technique for managing speech anxiety.

minutes. With these, you will be on your way to eliminating the sensation we call nervousness.

2.2   Primary Impact: Nonverbal Body Language

Studies indicate that 55% of the impact we make on an audience is communicated through our nonverbal body language. Body language is a broad term for forms of communication using body movements or gestures instead of, or in addition to, sounds, verbal language, or other forms of communication. A significant amount of the communication that goes on between people is nonverbal. Although most people do not realize it, and more cannot pick up on it, people are constantly using their bodies to send signs to each other. These signs can indicate what they are truly feeling at the time. Thus, reading body language can be useful in ascertaining exactly how others are feeling. The strong speaker is confident in her or his ability to overcome difficulties. This confidence will invariably translate to the quality of your demeanor and posture. This includes your eye movement, posture and body movement, gestures, and facial expressions.

2.2.1   Eye Contact

The eyes communicate powerful cognitive messages. You should not underestimate your ability to persuade an audience with your eyes.

Good eye contact communicates both personal confidence and respect for your audience. In addition, good eye contact helps your audience feel more relaxed and builds confidence in your speaking knowledge and ability. Maintain eye contact by knowing your speech well enough that you need only occasionally glance at your notes. Great speakers make a point of engaging their audiences by moving around the stage and even sometimes around the audience, ensuring that they make eye contact with everyone. Find a few friendly faces in the audience that react to your message and concentrate on delivering your speech to them. Keep eye contact for four to five seconds at a time, and then move to someone else. If you do not have the courage to make eye contact with your audience, then the audience will quickly lose interest in your speech. Think about the following questions:

- Do your eyes move slowly through the audience?
- Move rapidly?
- Move to the floor, ceiling, or walls?
- Do you lock eyes with one person in the audience before beginning to deliver the talk?
- Do you stay eye-to-eye with one person to finish a thought?

Remember, you want to be perceived as honest, sincere, confident, knowledgeable, and credible. What you do with your eyes plays an important role in how your audience perceives you, and how you look at people is more important than the fact that you look at them.

Walk to the front of the group at a deliberate pace. Look around the room silently. Take a few seconds to arrange your presentation materials. Breathe. Focus on a pair of eyes in the audience and hold gaze for a few seconds before you speak. Always focus on eyes before you begin speaking. Complete a phrase or thought with your eyes staying on one person. Focus on another pair of eyes before beginning your next thought. This small pause serves as punctuation in your talk, calms you, and gives your audience time to register each idea. Each phrase or thought should take between three and five seconds.

As you select people in different parts of the room, you will appear to be looking at more than one person at a time. This will give you control of the room and make the entire audience feel included. The "focus on eyes" techniques have the added advantages of controlling nervousness, reducing the number of filler words (see Section 2.3.4), and providing the presenter with audience feedback.

In summary, following are tips to achieve better eye contact with an audience:

- Mentally divide the room in three to five equal sections, and make eye contact with each section.
- When speaking from notes, follow these rules:
  - Look at your notes.
  - Absorb one idea.
  - Make eye contact with individuals in the audience.
  - Speak your full idea.
  - Repeat this pattern.

2.2.2   Posture

Good posture portrays confidence. We have all talked about the messages "body language" sends. Make sure that you have a proper posture. If your shoulders are sagging and your legs are crossed, you will not appear as being sincere, and people just will not accept your message. Square your feet shoulder-width apart, and plant your feet flat on the ground. Swaying or too much movement can be distracting to your listeners. Think about the following questions:

- Do your posture and stance command attention, without being stiff?
- Is your stance balanced? Are you pacing?
- Do you use your hands to describe or emphasize your point?
- Are your gestures above the waist?
- Where are your hands when not gesturing?
- Are your movements purposeful or random?
- Do your gestures give meaning to what you are saying?
- Are your gestures free-flowing?

Audiences see you before they hear you, and the way you stand and move is important in establishing a professional image. Maintain erect posture, a balanced stance, and face your audience. Following are tips for achieving better posture:

- Before your presentation, sit or stand straight.
- Walk confidently to the podium.
- Assume the "basic speaker stance," which is as follows:
  - Feet 12 to 18 inches apart and turned slightly outward
  - Weight evenly distributed over the balls of both feet
  - Body leaning slightly toward the audience
  - Hands at sides
  - Elbows loose

Relax your hands at your sides unless you are gesturing. To emphasize important points, you will want to use movements that are deliberate and precise, but look natural and spontaneous. Move purposefully.

2.2.3   Hand Gestures

Hand gestures are important to emphasize words and emotions, illustrate verbal messages, or even replace verbal messages altogether. Use gestures to paint a picture for the audience or underline key phrases. You should gesture crisply above the waist, so your audience can see your movements. Following are tips on using better hand gestures:

- Keep your hands by your side or neutrally in front of you, unless you can use them to make a point.
- Avoid doing the following with your hands:
  - Putting them behind your back
  - Putting them in your pockets
  - Putting them in front of you
  - Using them to fidget
- Keep most gestures at chest level.
- Think about what you are saying, and react with hand gestures to match or enhance your message.

2.2.4 Facial Expression

Pleasant facial expressions help to establish a warm and positive relationship with your audience. They enhance your verbal communication by producing feeling tone—the impression that you care about what you are saying. A smile lets them know that you are human and trustworthy, giving them more reason to accept your ideas. Expression in your face captures the mood of your speech and keeps the audience involved. Think about the following questions:

- What do you notice about your facial expression?
- Do you smile?
- Do your facial expressions convey enthusiasm? energy? excitement?

Most of us adopt a serious facial expression when asked to speak to a group. A smile is always appropriate and humanizes your talk. It relaxes you and your audience. It will be difficult to emphasize serious points in your talk if you maintain a serious facial expression throughout it. The following tips will help you use facial expressions to increase your persuasiveness:

- Smile before you begin speaking.
- Act naturally.
- Think about what you are saying, and react with facial expressions to match or enhance the thought.

Practice your speech in front of a mirror to evaluate your presentation style and body movements, and adjust accordingly (see Figure 2.2.). Try many different ways to find a comfortable balance of gestures to use in front of an audience.

Practice your speech in front of a mirror.

This is good to do, because you must

concentrate more. You also get an idea of

how you look when speaking. Finally, if

you must refer to notes, it allows you to

practice eye contact with the audience.

FIGURE 2.2
Practice your speech in front of a mirror.

2.2.5   Plan and Rehearse Your Movements

In conclusion, except for specialized briefings, a speaker should not be static. Movement and gestures are vital to maintaining speaker and audience enthusiasm for the presentation. Following are tips on ensuring lively (but not irritating) movement:

- If you move about on the stage, make your movements purposeful. Use your movement to reinforce or emphasize a point. Use it in concert with gestures to draw the audience into your presentation.
- Never turn your back on the audience while you are speaking.
- Avoid mannerisms that distract your audience such as pacing or swaying, hands in pockets, or hands waving around.
- Before the presentation, eliminate distracting items, such as change or keys, in your pockets.
- Be aware of your habits, such as crossing your arms, leaning against a wall or the podium, or tapping a pen. This might also be distracting to your audience or indicate to them that you are uninterested or not confident.
- Be aware of all potential obstacles around the stage. An embarrassing fall or trip will kill your concentration.

2.3   Secondary Impact: Control Your Vocal Quality, Volume, and Pace

Studies show that 38% of the impact of our presentations comes from what the audience hears—not the words, but the vocal quality, volume, pace, and expression we project.

> Practice with someone in the room to help you find your best public speaking volume. Increase your volume to account for the size of the room, the size of the audience and acoustics. Make sure Jim, who is sitting in the back row, can hear you clearly.

FIGURE 2.3
"Hey Jim, what kind of soda did you want again?"

2.3.1   Volume

One of the first things that happen when speakers get nervous is that the volume of their voices is significantly decreased. Consciously work to project your voice loudly enough for the audience in the back row to hear you. Think about the following:

- Is your volume clear and distinct or are some sounds slurred, dropped, or overdone?
- Can your audience hear you? Can your voice be heard in the back of the room?
- Does your volume convey conviction and confidence? Or does it suggest boredom?
- Does your voice fade at the end of sentences or phrases?

To help with your volume, be sure to do the breathing exercises. Even the most technically sound presentation falls apart if the audience cannot hear you; and your audience will not believe you if your voice is weak. Increase your volume to account for the size of the room, the size of the audience, and acoustics.

Picture a day that is not a speech day. Your boss is throwing a party for your project team because you have all done such a great job this year. You are standing up at the front, in charge of distributing the cans of soda. Jim is sitting in the back row and you cannot remember if he wanted Mountain Dew® or Coke®. So you say, "Hey Jim, what kind of soda did you want again?" (see Figure 2.3). Your speech needs to be at this volume level.

2.3.2   Pace

When you get nervous, you often talk fast—very fast. Your subconscious has whispered to you that the sooner you finish, the sooner you will be able to dash back to the safety of your desk. This is a feeling you will consciously need to fight. For example,

- Is your pace rapid, slow, varied, uneven, or too even? Is it fluid or halting?
- Do you run out of breath or hear yourself sigh?

As you practice your speech, work to speak slightly slower than you would normally. This will help you counteract the speech day speed-up. Pause to pace yourself and gather air to project your voice. Your audience appreciates the short pauses to absorb the information. You will need these pauses to collect yourself and breathe. Go easy on your listeners—take a nice, steady conversational pace.

2.3.3   Vocal Expression

Does your vocal expression convey enthusiasm, conviction, excitement, anger, joy, or seriousness? Is it flat or monotone? Do you emphasize key words? Is there a smile in your voice? Vary the tone and pitch of your voice to "underline" the key phrases that you want your audience to remember. For a free lesson on how to use vocal expression, listen to the anchors on the evening news.

2.3.4   Fillers

Using excessive fillers while you speak is the most irritating speech habit. Fillers range from repetitious sounds, such as "uh," "um," and "er," to favorite catch words and phrases, such as "you know," "anyway," "all right," and "like." The problem with using fillers such as these when you speak is that they distract your listener—often to the point that she or he does not hear anything you say. Your message is entirely lost, obscured by the thicket of fillers surrounding it.

Speech fillers are insidious. If you are a person who uses fillers, you may not even be aware of the speech problem yourself. Fillers tend to become so embedded into our speech patterns that once you become aware that you are using them, you will have a very hard time trying *not* to say them.

We use filler words to fill the pauses in our talk. By using the "focus on eyes" technique, you will dramatically reduce the number of filler words you use, and your audience will appreciate the pauses. As you become more conscious of the way you speak and practice speaking without fillers, you will find your filler use decrease. As your speech fillers decrease, your listeners will:

- Form a better impression of you as you speak, thinking of you as an educated, knowledgeable person who is more trustworthy
- Be better able to focus on the message you are communicating, instead of being distracted by the way you are expressing yourself

---

2.4   Optimize Your Presentation Environment

Pay careful attention to these conditions, no matter how basic they may seem. The speaker or writer who ignores them is in trouble before the first word is spoken or written. The answers call for imagination: Step back from your own interests, and see things from other angles. To speak or write well, you do not need to agree with your audience and readers. What is essential is to see the world from their perspectives.

The atmosphere, layout, size, and general suitability of the room you use for your speech can directly affect your presentation. Although it is true that you will be unable to influence this in some situations, the presenter normally has the opportunity to change things in the environment. Factors that affect you and your audience are room size, light, temperature and ventilation, room layout, equipment, noise, access, and time of day.

2.4.1   Room Size

The room should be large enough to permit the audience to sit in comfort without being so large that they feel lost in the room. Awkward shapes can be distracting or tiring to the eyes.

2.4.2   Light

The amount of light should be enough to see and not excessive. If natural light is entering the room, it should be behind the audience so as not to be distracting or tiring to the eyes.

Make sure you know where the switches for the lights are as well as the dimming capabilities in the room. Brief whoever is operating the lights so he or she knows what lighting you will require at what stages of your presentation.

2.4.3   Temperature and Ventilation

The hotter the room, the more likely the audience will fall asleep. Equally, the colder the room, the more distracted they are likely to feel by their discomfort. Therefore, a comfortable mid temperature, about 70°F, should be the goal.

Ventilation is particularly important for health and presentation effectiveness. Remember clean indoor air laws in most states will prohibit smoking.

When selecting a room, make sure that the ventilation system does not make excessive noise. Know where the controls are for the system.

2.4.4 Room Layout

When selecting the appropriate layout for your presentation, consider whether your audience needs tables for note taking, and whether they need to discuss things among themselves. The most common types of layouts are:

Cinema Style—Rows of chairs, one behind the other, are most suitable for large audiences or for watching videos.

Classroom—Add tables between rows. This is suitable where information is being passed from the speaker to the audience and very little participation is required from them.

U or V Shaped—Tables placed in the shape of a U or a V, and chairs behind them are particularly suitable where presentations may be followed by group discussion.

Boardroom Style—Individuals are seated around a single table or several tables. This is useful when frequent interaction is required among audience members. However, visibility may be restricted with this type of layout.

2.4.5 Equipment

Determine, well in advance, your equipment and power requirements. You will need to make arrangements for such things as a microphone, television, VCR, overhead projector, slide projector, screen, or extension cords.

2.4.6 Noise

Many locations have their own peculiar noises. Find out what they are in advance so you can avoid distracting yourself and the audience. Always know who is in the next room.

2.4.7 Access

People should normally enter the room from the back to avoid the disruption of latecomers.

2.4.8 Time of Day

Your audience is more alert in the morning. Whenever possible, plan to give your presentation early in the day. You will also avoid the "running late" problems usually created during the course of the day. Avoid late morning, early afternoon, and late in the day because audiences tend to be tired or distracted at those times.

Bibliography

Bailey, E. P., *A Practical Guide for Business Speaking*. New York: Oxford University Press, 1992.

Booher, D. D. and NetLibrary, Inc., *Speak with Confidence: Powerful Presentations That Inform, Inspire, and Persuade*. New York and London: McGraw-Hill, 2003.

Cialdini, R. B., *Influence: The Psychology of Persuasion*. Rev. ed. New York: Morrow, 1993.

Daley, K. and Daley-Caravella, L., *Talk Your Way to the Top: How to Address Any Audience Like Your Career Depends on It*. New York: McGraw-Hill, 2003

Davidson, J. P., *The Complete Guide To Public Speaking*. Hoboken, NJ: John Wiley & Sons, 2003.

Donnellon, A., *Team Talk: The Power of Language in Team Dynamics*. Boston: Harvard Business School Press, 1996.

Holtz, S., *Corporate Conversations: A Guide to Crafting Effective and Appropriate Internal Communications*. New York: AMACOM, 2004.

Isaacs, W., *Dialogue and the Art Of Thinking Together: A Pioneering Approach to Communicating in Business and in Life*. New York: Currency, 1999.

Leeds, D., Mohn, K., and NetLibrary, Inc., *PowerSpeak: Engage, Inspire, and Stimulate Your Audience*. Updated ed. Franklin Lakes, NJ: Career Press, 2003.

Levine, R. V., *The Power of Persuasion: How We're Bought and Sold*. Hoboken, NJ: John Wiley & Sons, 2003.

McGinty, S. M., *Power Talk: Using Language to Build Authority and Influence*. New York: Warner Books, 2001.

Morgan, N., *Working the Room: How to Move People to Action through Audience-Centered Speaking*. Boston: Harvard Business School Press, 2003.

Munter, M., *Guide to Managerial Communication: Effective Business Writing and Speaking*. 4th ed. Upper Saddle River, NJ: Prentice Hall.

Orben, R., *Speaker's Handbook of Humor*. Springfield, MA: Merriam-Webster, 2000.

Patterson, K., *Crucial Conversations: Tools for Talking When Stakes Are High*. New York: McGraw-Hill, 2002.

Reimold, C. and Reimold, P., *Short Road to Great Presentations: How to Reach Any Audience through Focused Preparation, Inspired Delivery, and Smart Use of Technology*. New York: John Wiley, 2003.

Simmons, J. *We, Me, Them, & It: The Powers of Words in Business*. New York; London: Texere, 2002.

Sparks, S., *Schaum's Quick Guide to Great Presentation Skills*. New York: McGraw-Hill, 1999.

Tingley, J. C., *The Power of Indirect Influence*. New York: AMACOM, 2001.

Urech, E., *Speaking Globally: Effective Presentations across Cultural Boundaries*. London: Kogan Page, 1998.

Valenti, J. *Speak up with Confidence: How to Prepare, Learn, and Deliver Effective Speeches*. New York: Hyperion, 2002.

Van der Heijden, K., *Scenarios: The Art of Strategic Conversation*. Chichester, England and New York: John Wiley & Sons, 1996.

Weissman, J., *Presenting to Win: Persuade Your Audience Every Time*. Upper Saddle River, NJ: Financial Times/Prentice Hall, 2003.

Whalen, D. J., *I See What You Mean: Persuasive Business Communication*. Thousand Oaks, CA: Sage Publications, 1996.

Wiener, V., *Power Communications: Positioning Yourself for High Visibility*. New York: New York University Press, 1994.

3

Presentation Aids

3.1 Engineering: The Real da Vinci Code

Visual thinking in engineering can be inspiring. Leonardo da Vinci, the talented Italian Renaissance painter, draftsman, sculptor, architect, and engineer used it to find inspiration by looking at stains on the wall:

> I cannot forbear to mention ... a new device for study ... which may seem trivial and almost ludicrous ... [but] is extremely useful in arousing the mind ... Look at a wall spotted with stains, or with a mixture of stones ... you may discover a resemblance to landscapes ... battles with figures in action ... strange faces and costumes ... and an endless variety of objects ...

da Vinci was a "Quintessential Engineer,"—one who is able to effectively combine scientific intellect with artistic creativity. Perhaps best known as an artist, his name inevitably brings forth images of The Last Supper and Mona Lisa. He was also an architect and a scientist. Yet, if asked, he would ultimately refer to himself as an engineer. Driven by an unrelenting curiosity and an insatiable hunger for knowledge, da Vinci was an incredibly innovative thinker who perceived the world not only as his personal playground, but also as one with unlimited possibilities. From his fertile mind sprang designs of flying machines and instruments of war, as well as practical theories and concepts in engineering, mathematics, and science—many of which were centuries ahead of their times. He was among the originators of the science of hydraulics and probably devised the hydrometer; his scheme for the canalization of rivers still has practical value. He invented a large number of ingenious machines, including an underwater diving suit. Although not practicable, his flying devices embodied sound principles of aerodynamics. Leonardo recognized that levers and gears, when applied properly, could accomplish astonishing tasks. Gears were at the heart of nearly all his inventions, from the crane to the helicopter to the automatic turnspit.

Engineering visual thinking refers to a group of generative skills that, when practiced with rigorous discipline, results in the production of novel

> *Levers and gears were at the*
>
> *hearts of Da Vinci's inventions.*
>
> *Engineering visual thinking*
>
> *resulted in the production of*
>
> *novel and original ideas.*

FIGURE 3.1
Engineering inventions and visual thinking.

and original design ideas (see Figure 3.1). By seeking to discover visual forms that fit his or her underlying engineering experience, the practitioner of visual thinking comes to know the world's need for a new product and process. This practice of thinking with images alone is stressed to balance the overemphasis on verbal reasoning in current engineering communication. Visual thinking is high-order critical thinking conducted by inventive imagination. Picture thinking involves different categorization than verbal or linguistic processing. Linguistic thinking involves categorization of thought in defined, linear forms. It is serial, and it concentrates on detailed parts in the stimulus. Visual thinking involves categorization that is parallel and holistic.

Picture thinking, also known as visual thinking or visual/spatial learning, is the phenomenon of thinking through visual processing, where most people would think with linguistic or verbal processing. It is nonlinear and often has the nature of a computer simulation, in the sense that much data is put through a process to yield insight into complex systems, which would be impossible through language alone.

Thinking visually is often associated with the right half of the brain. The visual–spatial learner model is based on the newest discoveries in brain research about the different functions of the hemispheres. The left hemisphere is sequential, analytical, and time-oriented. The right hemisphere perceives the whole, synthesizes, and apprehends movement in space. Picture thinking could be called "nonlinguistic thinking," and people who do such information processing could be called "visual thinkers." It involves thinking beyond the definitions of language and has many personal referents to meaning that cannot be translated.

Engineers communicate with far more than words. Becoming adept at visual thinking increases your power and persuasiveness during business communication. How can you ensure that the messages you communicate visually increase your credibility, authority, and persuasiveness in the minds of listeners? The following guidelines will help you to become a visual speaker.

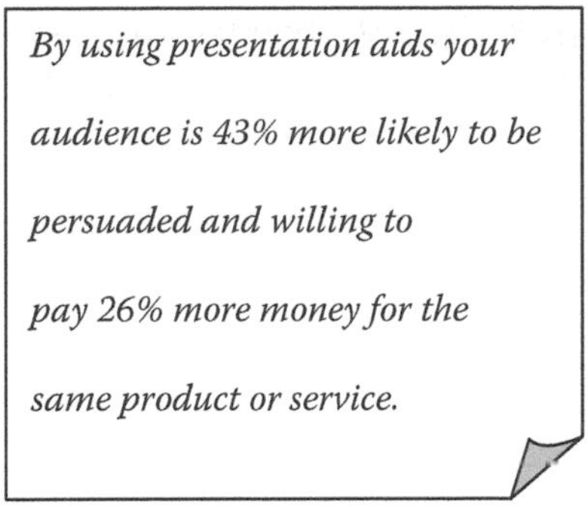

> *By using presentation aids your audience is 43% more likely to be persuaded and willing to pay 26% more money for the same product or service.*

FIGURE 3.2
Using presentation aids.

3.2 Speaking Visually—Guidelines for Using Presentation Aids

How effective are you at persuading audiences to see things your way? If you fail to speak visually, chances are the answer is "not very." Speeches benefit greatly from using presentation aids effectively. Just as a sentence in capital red letters stands out from all others, a speech with carefully planned and prepared aids will stand out to the audience. When you use presentation aids to support your message, you are perceived as more professional, more persuasive, more credible, more interesting, and better prepared. In addition, research on engineering presentation indicates that when you support your presentation with visuals, learning is improved by 200%, retention by 38%, and the time to explain complex business or technical subjects is reduced by 25% to 40%.

Studies indicate that if you stand up and give a presentation using visual aids, your audience is 43% more likely to be persuaded, and they are willing to pay 26% more money for the same product or service. Therefore, it makes sense to use visuals in your presentations.

Why don't more presenters use presentation aids? It requires some planning. Simply copying a page of engineering text onto an overhead does not achieve the results that studies have cited. When planning for presentation aids, it is important that you think of those aspects of your speech that may be difficult for the audience to understand from words alone (see Figure 3.2). For example, most people find it difficult to remember numbers, or they may have trouble visualizing items or process from your oral description.

One word of caution: You can use too many aids. Not becoming overwhelmed with the supportive visuals also requires some finesse. You can have so many visual aids that they actually detract from the message by distracting the audience.

An aid is not something that you throw together at the last minute simply because your instructor requires that you use one in your speech. Here is a checklist for using presentation aids.

____ Will my presentation aid enhance understanding?

____ Will a presentation aid help the audience remember the material?

____ Is my presentation aid easy to understand?

____ Is there enough information on my presentation aid?

____ Is there too much information on my presentation aid?

____ Is my presentation aid neat?

____ Is the print on my presentation aid large enough for all audience members to read?

____ Is everything on my presentation aid drawn to scale?

____ Do I have the necessary equipment to use my presentation aid?

____ Do I know how to use the equipment?

____ Will I need tape or thumbtacks to position my presentation aid?

____ Have I practiced presenting my speech using my presentation aid?

____ Could I give my speech just as well, if not better, without my presentation aid?

Regardless of the type of presentation aid you choose to use, remember three key concepts when preparing supplementary materials for a speech. Presentation aids should be:

- Big—An aid should be big enough that all members of the audience can see it easily without straining to see (or loud enough that each member of the audience can hear it easily). This includes any lettering or pictures that are included on the aid.

- Simple—Simplicity in terms of a presentation aid encompasses two different aspects of the aid. First, the aid should contain only relevant information. For example, instead of writing complete sentences, a speaker would want to put only key words or phrases on the aid. He or she could then explain the meaning of those words and phrases during the speech. In addition, simplicity refers to the design of the aid. Generally, it is better to use clear, uncluttered fonts, such as Times Roman or Helvetica, and stick to a simple design and color scheme.

- Attractive—Although presentation aids should play an informative function in your speech, they should also be interesting and attractive. Use balance and color. Incorporate clip art into aids that are all text. In addition, attractiveness means neatness. In most cases, presentation aids do not have to be of professional quality; however, lines should be straight, lettering should be neat, and words should be spelled correctly.

- *Flip charts do not need electricity.*

- *Flip charts are economical.*

- *Color can be added very easily.*

- *Flip charts allow spontaneity.*

FIGURE 3.3
Advantages of using a flip chart.

3.3   Choosing among Options

Commonly used ways of displaying visuals are flip charts, whiteboards, overhead projector transparencies (overheads), slides, computer presentations (e.g., PowerPoint), videotapes, and multimedia computer presentations.

Here are some guidelines for choosing visuals:

Flip Charts (see Figure 3.3) or Whiteboards:
- For display of sketches; charts, graphs, diagrams, photographs, artwork, or computer-generated images
- For using color contrast to your advantage
- For a maximum audience size of about 20
- For informal presentations
- Minimal production time
- Inexpensive
- Flexible
- Useful for recording audience input when brainstorming

Overheads:
- For maximum audience size of about 30, unless compact seating is available
- For formal or informal presentations
- Minimal production time
- Inexpensive
- Offer random access during the presentation
- Make sure images are clear and text is large enough to read.
- Do not overload an overhead.

Slides:
- For large audiences (several hundred)
- For formal presentations

- Require design and production time
- Moderately expensive
- Achieve photographic exactness
- For display of many visuals in rapid succession

Videotape:

- No more than 20 people per 21-inch television
- Larger audience capability with video projection equipment
- For formal or informal presentations
- When continuous interaction is not required
- Useful for showing role-plays or staged scenes
- Expensive to produce

PowerPoint:

- Can be used for large audiences
- For formal or informal presentations
- Minimal production time
- Relatively inexpensive (requires computer and software)
- Some flexibility in accessing visuals during speaking
- Allows you to create animated color images with sound
- Requires knowledge of how to work all equipment

Multimedia Presentations:

- Can be used for large audiences
- For formal presentations
- Require design and production time
- Expensive
- Easy to customize and highly flexible

3.4   Creating Visuals with Impact

Visual aids are vehicles for enhancing or for the further understanding of your spoken words. If they do not fulfill these purposes, they are misused. Design your visual aids to have a purpose or support an idea. Make sure your materials are large, clear, and uncluttered, and practice using them. Make sure you know how to turn all equipment off and on, and consider using an assistant to handle your aids so you can concentrate on your audience.

A visual becomes the focal point for the time it is in view. An audience's attention is quite naturally drawn from the speaker to anything that is put on a screen, blackboard (or whiteboard), or flip chart. The key to creating effective visuals is simplicity. Because they automatically assume center stage, it is vitally important that all visual aids support your talk in an attractive, comprehensible manner. Otherwise, they will detract from it and compete with it. Another issue to keep in mind is that, for many audience members, a darkened room is an invitation to doze. It is much easier to lose your audience when the lights are low, so be sure your visual material is compelling. You can support your talk with several types of visual aids, including charts and graphs, text visuals, and graphic visuals (pictures). In addition, color is 50% to 85% more effective in selling products and ideas. It accelerates learning, retention, and recall by 55% to 78%, and improves comprehension by up to 73%.

Charts and Graphs—Engineers can use diagrams and plots to display large amounts of information in ways that are easy to understand and help reveal relationships and patterns. When using charts and graphs, limit the data to only what is absolutely necessary. All numbers should be rounded to the nearest whole number, and all lines and axes should be clearly labeled. Add emphasis with color. Always check your accuracy when using visuals.

- *Line Graphs* show changes in variables over time.
- *Bar Charts* compare variables at one time or several points in time.
- *Divided Bar Charts* compare variables and their components at fixed times.
- *Pie Charts* show relationships to each other or a whole.
- *Maps* show geographical relationships.
- *Flowcharts* show process relationships.

Text Visuals—Text visuals use bullet points to support your message. Include only key words and not complete sentences. Keep visuals simple, with no more than three lines per visual and three words per line. Use large upper and lower case letters and emphasize visuals with color.

Graphic Visuals—More than any other presentation aid, graphics, pictures, cartoons, or drawings convey more information and ideas in less time to your audience. If one picture is worth a thousand words, then it is worth 8 minutes of talking, because we speak about 120 words per minute during a presentation.

Diagrams—Diagrams show part of a process, structure, or unit. Systems diagrams can help you visualize the links between parts of a system (e.g., major engine parts or the principle of sailing in equilibrium). Replace words with pictures, and use color to highlight major and minor links.

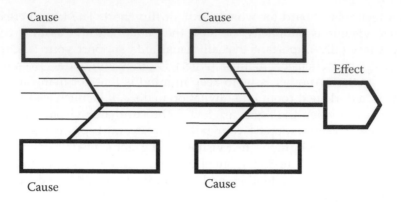

FIGURE 3.4
The fishbone diagram helps engineering problem solving.

The fishbone diagram is an analysis tool that provides a systematic way of looking at effects and the causes that create or contribute to those effects. Because of the function of the fishbone diagram, it may be referred to as a cause-and-effect diagram. The design of the diagram looks much like the skeleton of a fish. Therefore, it is often referred to as the fishbone diagram. The fishbone diagram is used to explore all the potential or real causes (or inputs) that result in a single effect (or output). Causes are arranged according to their level of importance or detail, resulting in a depiction of relationships and hierarchy of events. This can help you search for root causes, identify areas where problems may arise, and compare the relative importance of different causes.

Fishbone diagrams (see Figure 3.4) are frequently arranged into four major categories. Although these categories can be anything, you will often see:

- Manpower, methods, materials, and machinery (recommended for manufacturing engineering)
- Equipment, policies, procedures, and people (recommended for service engineering)

Whether you are using a poster or a computer-generated presentation, pre-planning and practice are important to effective use of presentation aids. Once you have your presentation aid prepared, practice your speech using the aid. This will help avoid the awkward pauses that can happen when the speaker is trying to decide how to put the poster up on the easel or how to hold the model while speaking. Practice will also alert you to potential problem areas, such as the need for a pointer or whether glare occurs on a screen. If you are using electronic equipment, be sure to check whether it is in working order. Consider the following:

- Are the necessary cords attached in the right places?
- Does the room where you will be giving the speech have an outlet available?
- Do you know how to use the equipment?
- If not, can you arrange for someone else to run it for you?
- Finally, have a backup plan in case something goes wrong.

3.5 Delivering with Visuals

Many people put their presentation aids up for audience scrutiny just before they begin speaking. However, that is not the most effective method for using a visual aid, because audience attention is focused on the newest, most interesting item—a presentation aid—instead of on the speaker. A better alternative is to follow a four-step plan: Introduce, Present, Explain, Put Away.

1. Introduce—Do not bring the presentation aid out for audience view until you are ready to discuss the point covered in the aid. This doubles the potential impact of the aid. For one thing, the audience is not used to it and tired of it already; for another, you get the added boost of attention at a point when you really want it.
2. Present—Once you do bring the aid out for audience view, give them a chance to look at it with their full attention. Usually, a pause of a second or two will give enough time for an audience to size up the general meaning of a presentation aid before you go on to explain what is included in the aid.
3. Explain—Occasionally, speakers will bring out attractive, well-planned presentation aids and then act as if they do not even know the aid is there. Use the presentation aid; do not simply display it. Refer directly to it; point to it if appropriate.
4. Put Away—Once you are finished with an aid, put it away (or turn it off) so that it does not continue to compete with you for attention.

It is easy to become overwhelmed with equipment and visuals during your presentation. The following techniques will help you stay in control and your audience gets the benefit of your message.

Sound-Bite Technique—Use the technique the television news professionals have made famous. Lead into your visual with an attention-grabbing statement that sets up your visual. In the next few seconds, explain the meaning and importance of your point using your visual as support. Close with a summary of your point and an introduction of what will come next in your presentation.

Ready-Aim-Fire—Maintaining one on one eye contact with your audience is especially challenging when you are presenting with visuals. Use the Ready-Aim-Fire technique to get your point across while keeping your "Focus on Eyes."

- Ready—Silently, look at your visual and think about what you want to communicate to your listeners. While you are looking at your visual, so are your listeners.

- Aim—Find a set of eyes in the audience.

- Fire—Talk to the set of eyes. Use the visual feedback you get from them to assess how your talk is being received. If they seem to be drifting off, pick up the pace. If they seem confused or unsure, slow down and repeat important points. Try to have your eyes on the audience 90% of the time you are speaking, particularly at the opening, the closing, and at the end of each emphasized statement.

Look back at your visuals as often as you need, but do not "talk to" the visuals. Remember that you should maintain control of the speaking situation. Do not distract from your speech with each presentation aid. In other words, be sure that you talk to the audience, not the presentation aid. Stand to the side of your aid so that the audience's view is not blocked, which is especially important when using an overhead projector.

With the help of the presentation aids, engineers see how ideas are connected, and they realize how information can be grouped and organized. With visual learning, new concepts are more thoroughly and easily understood when they are linked to prior knowledge. Linked verbal and visual information helps engineers make connections, understand relationships, and recall related details. Remember, even though you are using presentation aids, the emphasis is on your communication with the audience. If you can learn to create word pictures in listeners' minds, you have a much greater chance of getting your message across in a way that is both resonant and memorable.

Bibliography

Alley, M., *The Craft of Scientific Presentations*, New York: Springer-Verlag, 2003.

Bailey, E. P., *A Practical Guide for Business Speaking*. New York: Oxford University Press, 1992.

Booher, D. D. and NetLibrary, Inc., *Speak with Confidence: Powerful Presentations That Inform, Inspire, and Persuade*. New York and London: McGraw-Hill, 2003.

Cialdini, R. B., *Influence: The Psychology of Persuasion*. Rev. ed. New York: Morrow, 1993.

Daley, K. and Daley-Caravella, L., *Talk Your Way to the Top: How to Address Any Audience Like Your Career Depends on It*. New York: McGraw-Hill, 2003.

Davidson, J. P., *The Complete Guide to Public Speaking*. Hoboken, NJ: John Wiley & Sons, 2003.

Donnellon, A., *Team Talk: The Power of Language in Team Dynamics*. Boston: Harvard Business School Press, 1996.

Holtz, S., *Corporate Conversations: A Guide to Crafting Effective and Appropriate Internal Communications*. New York: AMACOM, 2004.

Isaacs, W., *Dialogue and the Art of Thinking Together: A Pioneering Approach to Communicating in Business and in Life*. New York: Currency, 1999.

Kenny, P., *A Handbook of Public Speaking for Scientists and Engineers*. Bristol, UK: Adam Hilger, Ltd., 1983.

Leeds, D., Mohn, K., and NetLibrary, Inc., *PowerSpeak: Engage, Inspire, and Stimulate Your Audience*. Updated ed. Franklin Lakes, NJ: Career Press, 2003.

Levine, R. V., *The Power of Persuasion: How We're Bought and Sold*. Hoboken, NJ: John Wiley & Sons, 2003.

McGinty, S. M., *Power Talk: Using Language to Build Authority and Influence*. New York: Warner Books, 2001.

Morgan, N., *Working the Room: How to Move People to Action through Audience-Centered Speaking*. Boston: Harvard Business School Press, 2003.

Munter, M., *Guide to Managerial Communication: Effective Business Writing and Speaking*. 4th ed. Upper Saddle River, NJ: Prentice Hall.

Orben, R., *Speaker's Handbook of Humor*. Springfield, MA: Merriam-Webster, 2000.

Patterson, K., *Crucial Conversations: Tools for Talking When Stakes are High*. New York: McGraw Hill, 2002.

Reimold, C. and Reimold, P., *Short Road to Great Presentations: How to Reach Any Audience through Focused Preparation, Inspired Delivery, and Smart Use of Technology*. New York: John Wiley, 2003.

Simmons, J., *We, Me, Them, & It: The Powers of Words in Business*. New York and London: Texere, 2002.

Sparks, S., *Schaum's Quick Guide to Great Presentation Skills*. New York: McGraw-Hill, 1999.

Tingley, J. C., *The Power of Indirect Influence*. New York: AMACOM, 2001.

Urech, E., *Speaking Globally: Effective Presentations across Cultural Boundaries*. London: Kogan Page, 1998.

Valenti, J., *Speak up with Confidence: How to Prepare, Learn, and Deliver Effective Speeches*. New York: Hyperion, 2002.

Van der Heijden, K., *Scenarios: The Art of Strategic Conversation*. Chichester, England and New York: John Wiley & Sons, 1996.

Weissman, J., *Presenting to Win: Persuade Your Audience Every Time*. Upper Saddle River, NJ: Financial Times/Prentice Hall, 2003.

Whalen, D. J., *I See What You Mean: Persuasive Business Communication*. Thousand Oaks, CA: Sage Publications, 1996.

Wiener, V., *Power Communications: Positioning Yourself for High Visibility*. New York: New York University Press, 1994.

4

Organize Your Talk

You have been asked to make a presentation because someone thinks you have something of value to say. How you present yourself determines the degree of credibility you achieve with your audience. What you say goes a long way in either persuading or turning off your listeners.

Your talk must address the emotional position of your audience as well as present the facts. How you organize your talk is based on the learning styles of the individuals in your audience as well as your objective. What you say depends on the knowledge and attitudes of your listeners.

4.1 Planning Your Talk

The first critical step in planning your talk is establishing objectives. In determining your objective, you should ask yourself, "Why am I making this presentation?" instead of "What am I going to say?" Consider all the following questions carefully in determining your objective.

- What reaction do you want from your audience?
- What action do you want your listeners to take as a result of your presentation?
- What are you trying to accomplish? (Decide what you wish to speak about.)
- Decide what is the primary purpose of the speech.
- Do you wish to:
  - instruct and inform?
  - convince, persuade, influence or motivate?
  - amuse and entertain?
- What are you trying to achieve?
- What are the objectives of your speech?
- Know your audience (see Section 4.2).

Think about your objective, and write it down on a piece of paper. This becomes the thread that will tie all aspects of your talk together. If you cannot say it in one sentence, you have too many objectives (see Figure 4.1).

> *Use the one-sentence objective as the thread that will tie all aspects of your talk together. Let the audience know how your presentation objective fits in with their professional goals.*

FIGURE 4.1
If you cannot say it in one sentence, you have too many objectives.

In one sentence, write down the objective of your speech. This sentence will become the criterion against which all material is judged for inclusion in your presentation. If a number of points need to be discussed, establish a theme (i.e., a central idea or concept that gives unity, direction, and coherence to the presentation as a whole). List the main points to be covered, and arrange them in a logical sequence.

4.2   Conducting an Audience Analysis: 39 Questions

When you determined your objective for the presentation, you focused what you wanted to accomplish in your talk. The next step considers your audience and what is most important to them. Knowing your audience helps you to select the type and amount of information to present and increases the probability of getting your point across.

This questionnaire is designed as a guide for knowing your audience. It will help you determine what to include in your talk and to identify the information that will be the most persuasive to them.

Know Who the Audience Is

1. Who asked you to give this talk?
2. What is the purpose of the meeting?
3. How many people will attend your presentation?
4. Who are they?
5. What is their relationship to you?
6. What is their relationship to your organization?
7. What do they have in common with each other?

8. How many are familiar with the subject?
9. How many equal or surpass your knowledge on the subject?
10. What are their attitudes, beliefs, and values on your subject?

Know What the Audience Wants

11. Why do they need to hear what you will say?
12. Will they have difficulty understanding you due to industry or professional jargon, references or names, places, events, products, or acronyms?
13. What is your credibility with this audience before you begin your talk?
14. What is it about you that could help your audience relate to you better?
15. What sincere compliment can you give this audience?
16. Where do audience members stand on your point of view?
17. What are the audience expectations of your talk?
18. What are the key decision makers in your audience?
19. What is the most important benefit to them?
20. Who is likely to support your point of view?

Know How to Appeal to the Audience

21. How can you appeal to them in your talk?
22. Which audience members are likely to have a negative point of view?
23. What are their concerns?
24. What facts do you have to address their concerns?
25. What challenges do you expect your audience to ask?
26. What time of day is your talk?
27. Where is your talk being given?
28. How formal/informal is the presentation environment?
29. How will the room be set up?
30. What special environment, lighting, or other needs do you anticipate?

Know How to Interact with the Audience

31. Are interruptions likely?
32. How long is your talk?
33. Is the program likely to be running late before you give your talk?
34. Who is speaking before you?
35. Who is speaking after you?

36. Are you being introduced? What are they likely to say?

37. Will you have to introduce another speaker/presenter?

38. Will other presentations given that day also support your point?

39. What other things do you know about your audience that are helpful to consider?

4.3   Organizing Your Talk in Seven Easy Stages

You have established an objective and you know your audience. By using effective organization skills you will help ensure you meet your objective. The audience will understand the value of your message and the actions they need to take.

Your audience is made up of individuals who have different learning styles, different personality styles, and varying degrees of interest in your presentation. The seven stages of organizing your talk are based on how engineers learn and how they are persuaded. Let us look at each stage separately and use these steps to develop your talk.

Stage 1 Open:

In this stage, you set the tone for the presentation by developing rapport, establishing credibility, and sharing your objective. Rapport is developed by making a connection with your listeners. This can be done through sincere compliments and sharing something about yourself that the audience can relate to. When you state your objective for the talk, it helps your audience understand the purpose of the presentation and how you are going to spend their time. Answering the following questions will help you prepare this stage of the presentation:

- What sincere compliment can you pay to your audience, their profession, or their organization?
- What can you share about yourself that your audience can relate to?
- What other things can you say to develop rapport?
- What can you say to establish your credibility with this audience?
- What will you say to establish the objective for your audience?
- What visual aids can you use to grab audience attention?

Stage 2 Explore:

In this stage, you explore the situation by discussing some key background information and relating it to your listeners' needs and interests. You will also explore any problems, or potential problems, related to your subject. Answering the following questions will help you prepare this stage of the presentation:

- What background information can you provide for your audience to help them understand your subject better?
- What can you say to ensure that the audience understands that your topic is valuable?
- What can you say to help them recognize the importance of your message if they are currently unaware of it?
- What problems is your proposal going to address?
- What potential problems will occur if your audience does not take the action you propose?
- What concerns does your audience have?
- What background information would be helpful to include on handouts or other take-away material?
- What visual aids can you use to make your point more clearly?

Stage 3 Explain:

Your idea or recommendation is what the audience is interested in hearing. You will want to outline your proposal and link it to the benefits and needs of your listeners. Present evidence to support your proposal, and connect it to the specific advantages of your idea. Avoid overwhelming them with too much information. They can absorb only so much, and you want to make sure it is your most important point that they remember. Answering the following questions will help you prepare this stage of the presentation:

- What is your idea or recommendation?
- What benefits will interest your audience?
- What evidence do you have to prove your point?
- What visual form should you use to present your evidence?
- What information would you like to include if you had more time? (Consider including this information in handouts.)

Stage 4 Invite:

In this stage, explain to your listeners what action they should take to implement your proposal. Be specific and brief, and continue to link the action to your main point and their needs. Answering the following questions will help you prepare this stage of the presentation:

- What do you want them to do?
- What is the timeframe?
- What visual aids can you use to encourage involvement?

Stage 5 Summarize:

This stage is not your close. This is a transition stage to lead you into encouraging audience response. Briefly summarize your objective, the situation, your proposal, and the actions you are requesting. Answering the following questions will help you prepare this stage of the presentation:

- How can you restate your objective?
- What can you say to tie your proposal to the situation?
- What can you say to encourage audience response?
- What visual aids can you use in this stage?

Stage 6 Request Reaction:

Now that you have shared your proposal with the audience, it is time to get their reaction. Do this by encouraging questions and comments. Use the "Handle with C.A.R.E." audience response technique for handling questions and concerns (see chapter 5, section 5.2). Answering the following questions will help you prepare this stage of the presentation:

- What can you say to encourage audience response?
- What challenging questions can you anticipate?
- What questions can you pose if no questions are asked?

Stage 7 Close:

This is the final stage of the presentation. It is the point at which you conclude audience questions and reinforce the main objective of your talk. Your audience will remember your close longer than any other part of your presentation. It must be strong, brief, and upbeat. Answering the following questions will help you prepare this stage of the presentation:

- How will you cue your audience to the close?
- What can you say to tie your presentation together and reinforce your main point?
- What can you say to close your presentation with power?
- What visual aid will be most memorable to your listeners?

4.4   Getting Attention and Keeping Interest

Your audience's attention and interest are critical for you to accomplish your objective. Your challenge is not only to make the audience want to listen, but also to help them understand, remember, and act on the information or ideas you share.

Nothing is more powerful in persuading your audience than your own commitment and enthusiasm for your topic. In addition to your own conviction, several techniques can be used to ensure that the impact you make is a positive one and that you hold their attention and interest throughout your presentation. These techniques include using questioning techniques, analogies, an inside story, numbers and statistics, hooks, handouts, audience participation, and humor.

4.4.1 Questioning Techniques

Your audience will remember less than 30% of the sentences they hear during your presentation. In comparison, they will remember more than 85% of the questions you ask. By asking questions, you deepen audience understanding and conviction.

The best questions are the ones that make your audience think, shock them to attention, or obtain their agreement. During your presentation, ask only questions resulting in a "yes" answer, a nod, or some form of agreement with your premise. Detailed questions that require audience input are better left to the question and answer (Q&A) session, so you can stay on track.

4.4.2 Analogies

You can get your point across in less time, with better understanding, and with longer retention if you use analogies. They are so effective that people will sometimes remember them forever. The more complex your subject, the more important it is to use analogies. Know your audience, however, because using a complex analogy to support complex material could be frustrating to your audience.

4.4.3 Inside Story

You will get your audience's immediate attention if you can give them the inside scoop on something, particularly when it hits close to home. Again, by knowing your audience, you can show them that they are receiving the most recent, relevant information available on the subject.

4.4.4 Personal Experience

Support the point you are making with first-hand experience. This not only enhances your credibility with the audience, but also proves your knowledge on the subject.

4.4.5   Startling Statistics

Numbers and statistics can lose your audience quicker than anything else. By handling numbers and statistics carefully, you can not only prove your point, but also surprise your audience. Present only the numbers and statistics that are necessary to make your point. Wherever possible, round to the nearest whole number. Graphs and charts should be simple, and detailed calculations should be provided on a handout.

4.4.6   Humor

Humor can be one of the most effective attention-getting techniques when used naturally and appropriately. Humor keeps the audience alert and awake. Laughter triggers the release of adrenaline and increases long-term retention of information. Humor makes audiences more relaxed, responsive, and creative.

It is important not to confuse humor with comedy. Nothing is more uncomfortable for an audience than a long, awkward, drawn-out, unnatural joke. Some of the funniest people never tell jokes. Many of them merely comment on how they interpret what is happening around them, and their unique perspectives are humorous.

4.4.7   Audience Participation

If people sit too long, their brains start to operate in a passive mode, not in an active mode conducive to learning, understanding, and retention. Get them involved in activities that help them get to know one another or help gather information. Games that relate to your subject are an excellent way to increase attention and interest. Encouraging your listeners to take notes will also help maintain their attention.

4.4.8   Hooks

Try using "hooks" throughout your presentation to increase intent. Arouse interest by giving your audience a clue about what to expect.

4.4.9   Handouts

Handouts are tools to supply your audience with complex or additional information to support your talk. Handouts are also tools for getting your audience involved. However, you will likely create a "heads-down audience" by distributing handouts at the beginning of your presentation. If not too disruptive, distribute the handouts as your audience needs them throughout your presentation.

4.5 "Five Minutes Early"—Time Management for Your Presentation

How many presentations have you attended that finish early? Finishing early requires good planning. Woodrow Wilson, the 28th President of the United States, said he would require 2 weeks to prepare for a 10-minute speech, 1 week for a 1-hour speech, but he could begin without notice to deliver a 2-hour speech.

A recent *Speakout* article (a publication of the National Speakers Association) indicated that 100% of people polled reported they dislike speakers going overtime. Therefore, it makes sense to finish early.

An inexperienced presenter will almost always run overtime. A good presenter will finish on time, but you should plan to finish five minutes early. By finishing your talk five minutes early, you will convey to your audience that their time is valuable, without depriving them of the material they came to hear.

Time of Day: Your audience is most alert first thing in the morning. By being scheduled earlier in the day, you can also avoid potential problems that might cause delays in your presentation.

Start on Time: Do not wait for latecomers unless it is absolutely necessary. By starting on time, you will set the tone for the rest of the presentation. Your listeners will see that you value their time as well as your own, and they will appreciate it. They are also more likely to return from breaks on time if you establish the importance of time management at the beginning of your presentation.

Plan to Finish Early: Plan to finish five minutes early. Be aware of time management from the very beginning of your presentation. Plan how long each stage of your talk should take, and stick to it.

4.6 Delivering Your Introduction

A good introduction has four major components:

1. Greet your audience.
2. Capture attention.
3. Establish your credibility.
4. Preview your talk.

4.6.1 Greet Your Audience

As with any conversation or interaction, a greeting is expected. This breaks the ice and establishes rapport with the audience.

Make sure that your greeting is not overly casual. Do not begin with, "Hi, I'm going to talk to you about … ". Better yet, choose a conventional greeting such as:

> "Good afternoon. My name is Bill Lee. I'm pleased to talk with you about a topic that is important to engineers: listening skills."

4.6.2 Capture Your Audience's Attention

Suggest an intriguing opening. You might use a pithy narrative, a compelling statistic, or a relevant quotation.

> "Let me ask you this: On a scale of 1 to 10, how would you rate yourself as a listener? [pause] Most U.S. adults would rate themselves a 7.5. Unfortunately, according to the National Communication Association, if you're like most adults, you listen with just 25% efficiency!"
>
> "If you're like me, you realize that there is more to listening than meets the ear," or " If you're like me, you know that it's in your best interests to improve your listening skills."

4.6.3 Establish Your Credibility

You want to ensure that your audience understands your expertise and experience in your topic. Otherwise, they are not likely to be very receptive to your sage words.

> "I've been studying listening and training people to improve their listening skills for more than 20 years," or "I spent 14 years training listeners for a suicide hotline as well as other professionals who recognize the importance of good listening."

4.6.4 Preview Your Talk

Orient your audience to the format and central idea for your talk.

> "For the next 30 minutes, I'd like to share with you the 5 important listening considerations that I believe will help you get along better with your family and friends, and your colleagues."
>
> "I'd like to present to you some research and practical tips for enhancing your listening skills."
>
> "With the hope that you'll be actively listening to me, I invite you to stop me at any time and to ask me for clarification. I promise to listen to you, too."

4.7 Presenting Your Conclusion

A good conclusion has four major components:

1. Signal the end of your talk.
2. Summarize your main points.
3. Suggest a call to action or provide a memorable statement.
4. Thank your audience for listening.

4.7.1 Signal the End of Your Talk

Audiences appreciate this gesture. Some will wake up, whereas others will begin to consider the questions that they might ask you should you offer a Q&A session. Still others will stop fidgeting. Occasionally, audience members will look at their watches in disbelief that the time has passed so quickly.

> "Before concluding my talk, let me remind you of the three most important elements of well-designed visual aids."

4.7.2 Summarize Your Main Points

Focus upon the key points that prove your thesis true. If you left something out during the talk, let it go. Once you have promised to conclude your remarks, do so!

> "As I hope you'll recall, I believe that effective visuals are big, bold, and brief."

4.7.3 Suggest a Call to Action or Provide a Memorable Statement

For persuasive or motivational talks, you want to give your audience specific steps to take next. An informative talk requires a pithy ending. Often, this is a quotation or the rest of the story (from your introductory attention getter).

> "Whether you are using newsprint, PowerPoint, or a whiteboard, your audience will appreciate it if you will K.I.S.S. your visuals. That is, Keep it Simple, Speaker."

4.7.4 Thank Your Audience for Listening

As with any conversation, you want to end your talk with a polite gesture that signals that you have concluded and allows your audience to start applauding.

> "Thank you for listening. I appreciate your kind attention."

Bibliography

Alley, M., *The Craft of Scientific Presentations*. New York: Springer-Verlag, 2003.

Bailey, E. P., *A Practical Guide for Business Speaking*. New York: Oxford University Press, 1992.

Booher, D. D. and NetLibrary, Inc., *Speak with Confidence: Powerful Presentations That Inform, Inspire, and Persuade*. New York and London: McGraw-Hill, 2003.

Cialdini, R. B., *Influence: The Psychology of Persuasion*. Rev. ed. New York: Morrow, 1993.

Daley, K. and Daley-Caravella, L., *Talk Your Way to the Top: How to Address Any Audience Like Your Career Depends on It*. New York: McGraw-Hill, 2003.

Davidson, J P., *The Complete Guide to Public Speaking*. Hoboken, NJ: John Wiley & Sons, 2003.

Donnellon, A., *Team Talk: The Power of Language in Team Dynamics*. Boston: Harvard Business School Press, 1996.

Holtz, S., *Corporate Conversations: A Guide to Crafting Effective and Appropriate Internal Communications*. New York: AMACOM, 2004.

Isaacs, W., *Dialogue and the Art of Thinking Together: A Pioneering Approach to Communicating in Business and in Life*. New York: Currency, 1999.

Kenny, P., *A Handbook of Public Speaking for Scientists and Engineers*. Bristol: Adam Hilger, Ltd., 1983.

Leeds, D., Mohn, K., and NetLibrary, Inc., *Powerspeak: Engage, Inspire, and Stimulate Your Audience*. Updated ed. Franklin Lakes, NJ: Career Press, 2003.

Levine, R. V., *The Power of Persuasion: How We're Bought and Sold*. Hoboken, NJ: John Wiley & Sons, 2003.

McGinty, S. M., *Power Talk: Using Language to Build Authority and Influence*. New York: Warner Books, 2001.

Morgan, N., *Working the Room: How to Move People to Action through Audience-Centered Speaking*. Boston: Harvard Business School Press, 2003.

Munter, M., *Guide to Managerial Communication: Effective Business Writing and Speaking*. 4th ed. Upper Saddle River, NJ: Prentice Hall.

Orben, R., *Speaker's Handbook of Humor*. Springfield, MA: Merriam-Webster, 2000.

Patterson, K., *Crucial Conversations: Tools for Talking When Stakes Are High*. New York: McGraw-Hill, 2002.

Reimold, C and Reimold, P., *Short Road to Great Presentations: How to Reach Any Audience through Focused Preparation, Inspired Delivery, and Smart Use of Technology*. New York: John Wiley, 2003.

Simmons, J., *We, Me, Them, & It: The Powers of Words in Business*. New York and London: Texere, 2002.

Sparks, S., *Schaum's Quick Guide to Great Presentation Skills*. New York: McGraw-Hill, 1999.

Tingley, J. C., *The Power of Indirect Influence*. New York: AMACOM, 2001.

Urech, E., *Speaking Globally: Effective Presentations across Cultural Boundaries*. London: Kogan Page, 1998.

Valenti, J., *Speak up with Confidence: How to Prepare, Learn, and Deliver Effective Speeches*. New York: Hyperion, 2002.

Van der Heijden, K., *Scenarios: The Art of Strategic Conversation*. Chichester, England and New York: John Wiley & Sons, 1996.

Weissman, J., *Presenting to Win: Persuade Your Audience Every Time*. Upper Saddle River, NJ: Financial Times/Prentice Hall, 2003.
Whalen, D. J., *I See What You Mean: Persuasive Business Communication*. Thousand Oaks, CA: Sage Publications, 1996.
Wiener, V., *Power Communications: Positioning Yourself for High Visibility*. New York: New York University Press, 1994.

5

Handling Audience Response

You have leaned how to build an effective presentation, with an opening, a body, and a conclusion, as well as how to incorporate gestures and vocal variety. Your presentation does not end once you have finished what you have to say. The question period often is the part of the talk that influences the audience the most. After all, you have had time to practice the rest of the talk. This is the part of the presentation where your ability to interact with the audience will be evaluated. Question and answer (Q&A) sessions are where presentations enter the real world. They are where the audience shifts from passive to active.

Handling questions can be nerve-wracking because potentially, you may get questions that you cannot or perhaps do not want to answer. However, just as with presentations, preparation is a vital tool to help you perform with ease and confidence in a Q&A session. Questions or comments from the audience, either during the presentation or following it, offer several opportunities. They can help to ensure your message has been received as intended, any misunderstandings are clarified, and important points are reinforced.

For the audience, such an interaction is an opportunity to question the validity of the material presented, collect additional details, and evaluate how well the speaker thinks on his or her feet. Making the most effective use of audience response requires considerable skill and planning because the impression you make can either make or break the presentation.

How can you prepare for something you cannot control? It may seem impossible. However, if you think about it, you do have a fair idea as to:

- The questions that might be raised
- The expected attitude of the audience (i.e., hostile, friendly, curious, or confused)

The best way to prepare yourself and build your confidence is to take the time to write down as many possible questions as you can think of, and then practice answering them before the event. To get more ideas of possible questions, you can ask others to pose questions to you and practice answering them. It is particularly important to practice answering what you consider the most difficult questions. Craft and rehearse the answers to these difficult questions before the presentation.

> Only interested participants will take the time to ask questions. Make it easy for participants to ask questions. Also, provide a phone number or an e-mail address they can use to ask questions after your presentation.

FIGURE 5.1
Encourage questions.

5.1 Create the Environment

Effective presenters will have established a climate of trust and respect from the very beginning of their presentations. They will have anticipated common questions and handled those in their presentations. Although they have encouraged participation throughout the presentations, effective presenters will have asked their audiences to hold questions for a Q&A session at the end. This helps the presenter maintain control, stay on track, and finish on time.

5.1.1 Encourage Questions

Always cue the audience early that you will be taking questions. Unless you are giving a formal speech to a large audience, be prepared to take questions throughout your talk, not just at the end. You may, however, want to save Q&A for specific times during your presentation. When you ask for questions, use a technique that assumes the audience has questions, such as "What questions do you have?" Be sure to wait a little while. It takes people a moment to reflect on what you have said (see Figure 5.1).

5.1.2 If No Questions

You might have sat through presentations where the presenter wraps up by saying, "Any questions?", fails to pause, and then responds with "I guess I must have answered all your questions, so I am done." An audience may not have questions for several reasons. The least likely reason is that you have already answered all potential questions. However, it more likely that

- They are afraid of looking stupid.
- They are tired or bored.

- They are confused.
- They are frustrated.
- They are angry.
- They want to leave.

Fortunately, most of these reasons can be overcome by good planning, knowing your audience, and creating a positive presentation environment. If there really are no questions, be prepared to ask your own. You might say, "An important question that often comes up is ... ", and then answer your own question. Others will follow.

5.2 Handle with C.A.R.E.

You have worked very hard to deliver a dynamic presentation. You were well prepared, covered all your material in the allotted time, used a flawless delivery style, and want to maintain the credibility you have worked hard to establish. Now you have to relinquish some control and open the floor to questions. What if you misunderstand a question, offend someone, get rattled, or are challenged? By handling the Q&A session with C.A.R.E., you can double the impact of your presentation and keep your credibility intact. For C.A.R.E., "C" means Clarify, "A" means Amplify, "R" means Respond, and "E" means Encourage.

5.2.1   C = Clarify

Listen to the entire question. Make sure you understand the question you are being asked. Repeat it in your own words to check out your understanding Listen carefully not only to what is being said, but also to the feelings. Check your understanding of the issue while maintaining eye contact with the questioner. If you do not hear the question or understand it, ask the questioner to repeat it (see Figure 5.2).

Listen to the entire question *before* you begin to answer. Too many people begin responding to a question before the entire question is even asked. By not waiting to hear the entire question, you might give a response that has nothing to do with the question. Force yourself to listen to the entire question, and make sure you understand the question.

5.2.2   A = Amplify

Involve everyone by glancing at someone other than the questioner. Restate the question loudly enough for everyone to hear, and rephrase the question clearly enough for everyone to understand. Repeating the question will

Did you get to the specific, detailed question?

Make sure to get specific details by clarifying

the question. To answer questions effectively,

get information about a particular point such as

how much, when, what projects, products, or

manufacturing processes. As Mark Twain said,

"It ain't what you don't know that gets you

into trouble. It's what you know for sure that

just ain't so."

FIGURE 5.2
Clarify questions.

allow you some additional time to evaluate it and formulate a response. It is important that everyone "hear" the question or else the answer you provide may not make sense to some of the audience members. This is especially important when you are being recorded. A microphone may not pick up audience questions. When "amplifying," use this opportunity to formulate your answer.

5.2.3   R = Respond

Answer to the best of your ability. Try to be brief. When responding to the question, phrase your answer so it is directly relevant to your presentation objectives. Be brief; the Q&A session is for your audience to collect more information, not for you to talk some more. Honesty is the best policy. If you do not know the answer to something, admit it—you can offer to get back to the person later with the answer.

Resist the temptation to maintain eye contact with only the questioner. End eye contact with the questioner if you see value in doing that. Otherwise, end with someone else.

5.2.4   E = Encourage

Encourage additional questions. Ask, "Does that answer your question?" and "Is that the kind of information you were looking for?" This is critical. Once the questioner responds, "Yes," you now have permission to go on to the next person. This also gives that person one more opportunity to say,

"No" and allows them to clarify their question more by asking it again. Give all audience members a chance to ask questions.

5.3 Deal with Hostile Questions

When a tough situation comes your way, everyone in the audience waits to see how you will react. Your ability to skillfully handle these questions will go a long way to enhancing your credibility.

Try to stay calm, even if your audience is hostile. Always respect the questioner, even if you do not like the question or the manner in which it is posed. Maintain a positive attitude at all times. Never say or do anything that suggests you feel it is a tough question. It is not as simple as the cliché: "Never let them see you sweat." However, it is a good starting point. If you look at each questioner and each question in a polite and pleasant manner, audience members will not feel like they got under your skin. Follow the next four steps for dealing with a hostile questioner.

5.3.1 Address Emotions

Listen carefully to the question, and repeat it aloud. Make sure you understood the question correctly and that your audience knows the question to which you are responding. Your body language must be totally consistent with your confident voice and answers. You must smile pleasantly, look directly at the questioner, and gesture normally. Avoid crossing your arms, shaking your head, or putting your arms on your hips in a defiant manner.

Too often, we jump in with a logical answer to an emotional statement. This only antagonizes the questioner. Instead, gain control of tough questions and lead the momentum to your positive information, or wherever you want to direct everyone's attention. To put the angry questioner at ease, acknowledge his or her emotions. Demonstrate that you understand by responding to what the person is feeling. An effective presentation creates an aura of confidence. When the questioner is hostile, respond as if he or she were a friend. Any attempt to "put down" your questioner with sarcasm will immediately draw the audience's sympathy to the questioner. The purpose is to show understanding, not agreement. Only after you have shown that you understand the questioner's emotional position will he or she be able to hear the logic of your response.

5.3.2 Separate Content from Tone and Restate

When dealing with a confrontational question, separate the attitude of the questioner from the content of the question. For example, consider this question:

"Why do we need to waste our time learning this? The old system works fine, and this one is full of bugs."

Separating the tone of this question from its content defuses this question. The tone is challenging. If you respond to the tone, with a challenging or sarcastic response, you decrease your credibility.

However, the content is a legitimate question, and can be rephrased and restated by you in a less confrontational way. For example, you could restate the question:

"We have a question about what advantages this new system has over the old one, and, what do we do when we encounter bugs in the new system?"

5.3.3 Address Issues

Neutralize the question as you rephrase it for the entire group. Do not use the questioner's negative language. Never change the issue in the rephrasing. Give simple answers to simple questions. If the question demands a lengthy reply, agree to discuss it later with anyone interested. Whenever possible, tie your answer to a point in your speech. Look upon these questions as a way to reinforce and clarify your presentation. Always tell the truth. If you try to bend the truth, you almost always will be caught. Play it straight, even if your position is momentarily weakened.

The best way to handle tough questions from a tough audience is to pretend that neither exists. There is no such thing as a tough question. There can be easy questions where you know the answer or tough questions for which you do not have an answer. In that case, the answer is easy because all you have to do is say, "I don't know." Then, announce that you will find out the answer if it is ascertainable. If the question is inherently unknowable, simply say you do not know and move on.

5.3.4 Address the Audience

Once the questioner sees that you really understand both the feelings and the issues of the question, turn physically and move toward the whole audience, signaling the question is for everyone—then respond. Avoid a one-on-one dialogue with the questioner. When listening to the question, look at the questioner while moving away to include the whole group. Paraphrase the question for the group. State your answer to the group. Beware of answering only to the questioner.

Finish your eye contact with someone other than the questioner. A good Q&A exchange has a certain rhythm. It volleys back and forth in a brisk manner. Keep your answers brief and to the point, with many members of the audience participating. Be prepared with some appropriate closing remarks, and end with a summary statement that wraps up the essential message you want them to remember.

5.4 Deal with Other Types of Questions

5.4.1 Dealing with Rambling Questions

The rambling question goes on and on until the listener is unsure what is being asked. When faced with a rambling question, rephrase and simplify the question to help refocus the discussion for others. Many questioners ramble, and it takes some skill to identify the main issue. If needed, offer to clarify and address it after the lecture.

5.4.2 When You Have No Answers

You may not be able to answer a question for many reasons—many of them quite understandable. Try these ideas:

- Say that you do not know, but offer to find the answer later.
- Say that you do not know, and suggest a source where the questioner might find the answer.
- Say that you do not know, but answer a related question.
- Try the boomerang technique: "I don't know, what do you think?"
- Defer the question to someone else in the room—a person who might know the answer. If you display confidence, they will handle the question professionally. Remember, your credibility is still on the line.
- Invite the questioner to speak with you after the presentation when you can write his or her question down.
- Have the questioner write the question on the back of his or her business card.
- If you are compelled to answer and are not absolutely sure of your facts, state this to your listeners.

Admitting that you have no answers directly provides a good opportunity to ask for audience opinion and for interaction with questioners later (see Figure 5.3). Mark Twain described the following prompt answer from a steamboat pilot in *Life on the Mississippi*:

> I was gratified to be able to answer promptly, and I did. I said I didn't know.

> If you don't know the answer to something, admit it. You can offer to get back to the person later with the answer. Remember what Mark Twain said, "It ain't what you don't know that gets you into trouble. It's what you know for sure that just ain't so."

FIGURE 5.3
"I said I didn't know."

5.5 Control the Q&A Session

Q&A sessions are great opportunities to demonstrate your confidence, your sense of humor, and get audience participation during a public speaking engagement. To prepare for Q&A sessions, you should spend some time anticipating questions.

5.5.1 Manage Your Time

In designing your presentation, include a specified time for questions. Explain at which points during the presentation you will take questions. Formally recognize the questioner before he or she speaks, and limit the number of questions. Be sure to let your audience know how much time you have allowed for questions. This will give you the freedom later to tell them you only have time for one or two more questions. No matter how exciting your presentation is, always let your audience leave on time.

5.5.2 Wrapping Up

Never end your presentation with a question period and close by saying, "No more questions? Well that's all." This is a weak closing. Always save an upbeat, memorable piece to conclude your talk. Let the audience leave with your final thought instead of whatever the last question happened to be.

Although a Q&A session can be stressful, it can offer you an opportunity to clarify things your audience may not have understood; repeat things you think are important. In ending the session, you will usually have the last word. Use it to summarize your position or stress what you think is the most important point of your presentation. This will be your last chance to impress or inform your audience, so use it to your advantage.

5.6 Thinking on Your Feet

As you may have discovered, speaking accurately, fluently, and flexibly, especially under pressure, can be difficult. You may be prepared to present a proposal or sell an idea, but unless you can quickly translate your ideas into speech and language, your message will be undermined.

During these times, we can feel the pressure. Our hearts begin to race, we start to sweat, we feel our knees knocking, or we want to hide under a rock. This is because our delivery and response to questions will affect our objective, whether it is funding for a large project, customer satisfaction, a project leadership opportunity, or presentation success. People who can think and speak on their feet solve problems faster. They are perceived as more confident and creative, more persuasive, and trustworthy. Techniques that will help you quickly obtain appropriate speech and language for everyday use in your work and personal life are described next. The six secrets will help you master your "thinking on your feet" skills.

5.6.1 Listen

Many times when we are in high-pressure situations, we become so nervous that we really do not hear the actual question. We have all "Been there, done that." To make sure you understand the question and give the right answer, do the following:

- Breathe slower (Benefit: relaxes body and mind.)
- Look directly at the questioner. (Benefit: increases comprehension.)
- Ask questions (Benefit: increases clarity and shows you are listening.)

Slow down your body movements. If you have to rise from a seated position, do it slowly. Take in a slow two-second breath before you speak. While doing these "take time" techniques, plan internally how you will structure your speaking. By slowing down your body movements, you will help your mind to become calmer and better able to access thoughts and ideas.

5.6.2 Pause to Organize

Pause to gather your thoughts; a pause is acceptable. When you pause, you look and sound poised and in control. Remember, there is power in silence.

When you begin to speak, do so at a slower speed and use shorter words until you are speaking fluently. For self-trust in a difficult moment, silently say to yourself, "I can do this," or something similar that is a positive reinforcement of your ability.

5.6.3   Repeat the Question

You can repeat part of the question that an audience member might ask you. For example, if someone asks, "What are the key components of this system?" you can say, "Joe, the key components of this system are ... ." This has several benefits:

- Buys you time to think and communicate a complete piece of information: When you know your listeners likely will be biased against what you have to say, admit the situation—not bluntly, of course, but diplomatically. Chances are good that your listeners will give you a fair hearing. It is amazing what people will do when you simply acknowledge and ask.

- Allows you to take control of the question by rephrasing it in a more positive light if needed: Ask yourself if the question is appropriate. If a question is not relevant at the time, tell the person you will cover it later or on a one-on-one basis after the presentation. Make sure you do this clearly or you will lose credibility.

- Enables everyone, if in a public setting, to hear the question: Before responding, ask other members of the audience if they would like to respond. By being observant, you can identify people in the audience who can help you answer the question. While another person is giving his or her view, you have time to formulate your thoughts.

5.6.4   Focus on One Main Point and Support It

The number one reason why we sometimes freeze up when asked to think on our feet is that we have so many ideas running through our minds. We do not know which idea to mention first.

Here is the solution: Go with the first idea that comes to mind and say it. By sticking with that one point, you can focus on two or three supporting points. You sound more direct and confident when giving your answer.

5.6.5   Summarize and Stop (SAS)

End your answer with some SAS. Give your answer, summarize, and stop. Do not apologize, and do not ramble on beyond the finish. Try this trick: Repeat the essence of the question. For example, you may be asked, "Why did you stop the project?" In your summary, you might say, "And that's why we decided to start another project." Stop there.

When answering questions, let the audience know the end is near by saying:

"In summary ... "
"In conclusion ... "

Then, simply stop there. Remember SAS.

Apply these techniques so that you can become a master at "thinking on your feet" and give great answers. Have people fire questions at you in rapid succession and use the preceding strategies. The way to become proficient at thinking on your feet is by giving yourself many practice attempts. Subsequently, when the real situation occurs, you will have trained your mind and mouth to respond.

5.6.6 The Q&A Slide

This is the slide that follows your conclusions and remains in the background as you answer questions from the audience. Carefully select the most important images from the previous slides in your presentation, shrink them so they all fit on the last slide, and continue to project that slide for the audience to see as the Q&A session continues. This will allow the audience to consider your data and interpretations without having to recall details. It gives them a chance to reconsider the information after you have delivered the "big question" answer, and it helps you to guide the content of the Q&A session to meet your own objectives.

Bibliography

Alley, M., *The Craft of Scientific Presentations*. New York: Springer-Verlag, 2003.

Bailey, E. P., *A Practical Guide for Business Speaking*. New York: Oxford University Press, 1992.

Booher, D. D. and NetLibrary, Inc., *Speak with Confidence: Powerful Presentations That Inform, Inspire, and Persuade*. New York and London: McGraw-Hill, 2003.

Cialdini, R. B., *Influence: The Psychology of Persuasion*. Rev. ed. New York: Morrow, 1993.

Daley, K. and Daley-Caravella, L., *Talk Your Way to the Top: How to Address Any Audience Like Your Career Depends on It*. New York: McGraw-Hill, 2003.

Davidson, J. P., *The Complete Guide to Public Speaking*. Hoboken, NJ: John Wiley & Sons, 2003.

Donnellon, A., *Team Talk: The Power of Language in Team Dynamics*. Boston: Harvard Business School Press, 1996.

Holtz, S., *Corporate Conversations: A Guide to Crafting Effective and Appropriate Internal Communications*. New York: AMACOM, 2004.

Isaacs, W., *Dialogue and the Art of Thinking Together: A Pioneering Approach to Communicating in Business and in Life*. New York: Currency, 1999.

Kenny, P., *A Handbook of Public Speaking for Scientists and Engineers*. Bristol, UK: Adam Hilger, Ltd., 1983.

Leeds, D., Mohn, K., and NetLibrary, Inc., *PowerSpeak: Engage, Inspire, and Stimulate Your Audience*. Updated ed. Franklin Lakes, NJ: Career Press, 2003.

Levine, R. V., *The Power of Persuasion: How We're Bought and Sold*. Hoboken, NJ: John Wiley & Sons, 2003.

McGinty, S. M., *Power Talk: Using Language to Build Authority and Influence*. New York: Warner Books, 2001.

Morgan, N., *Working the Room: How to Move People to Action through Audience-Centered Speaking*. Boston: Harvard Business School Press, 2003.

Munter, M., *Guide to Managerial Communication: Effective Business Writing and Speaking*. 4th ed. Upper Saddle River, NJ: Prentice Hall.

Orben, R., *Speaker's Handbook of Humor*. Springfield, MA: Merriam-Webster, 2000.

Patterson, K., *Crucial Conversations: Tools for Talking When Stakes Are High*. New York: McGraw-Hill, 2002.

Reimold, C. and Reimold, P., *Short Road to Great Presentations: How to Reach Any Audience through Focused Preparation, Inspired Delivery, and Smart Use of Technology*. New York: John Wiley, 2003.

Simmons, J., *We, Me, Them, & It: The Powers of Words in Business*. New York and London: Texere, 2002.

Sparks, S., *Schaum's Quick Guide to Great Presentation Skills*. New York: McGraw-Hill, 1999.

Tingley, J. C., *The Power of Indirect Influence*. New York: AMACOM, 2001.

Urech, E., *Speaking Globally: Effective Presentations across Cultural Boundaries*. London: Kogan Page, 1998.

Valenti, J., *Speak up with Confidence: How to Prepare, Learn, and Deliver Effective Speeches*. New York: Hyperion, 2002.

Van der Heijden, K., *Scenarios: The Art of Strategic Conversation*. Chichester, England and New York: John Wiley & Sons, 1996.

Weissman, J., *Presenting to Win: Persuade Your Audience Every Time*. Upper Saddle River, NJ: Financial Times/Prentice Hall, 2003.

Whalen, D. J., *I see what you mean: persuasive business communication*. Thousand Oaks, CA: Sage Publications, 1996.

Wiener, V., *Power communications: positioning yourself for high visibility*. New York: New York University Press, 1994.

2

Write Your Way for Business Impact

6

Organizing for Emphasis

Weak writing takes too long to get to the point. It puts the "bottom line" at the bottom, in mystery-story fashion. The delay may work when the goal is to entertain casual readers, but at the office or on a factory floor, where we write to inform or persuade busy people, a slow buildup is inefficient.

Your readers need the most important information early to appreciate the relevance of anything else you say. If you postpone your "So what?", they must read your writing twice—once to see where the details are headed and again with the details in perspective.

Most of the time, get right to the point, in the same way newspaper articles do. These articles start with the most important information and taper off to the least important. Here is what to put first:

- Requests before justifications
- Answers before explanations
- Issues before background
- Conclusions before analyses
- Generalities before details

6.1 Make Your Bottom Line the Top Line

Correspondence of a page or less usually includes one sentence that is more important than any other. It is the one sentence you would keep if you could keep only one (see Figure 6.1). It is likely to be the one that tells readers what you want them to think or do. In timid writing, it appears at the end, after a terminal (i.e., read deadening) transition such as "therefore."

That key sentence should appear early—by the end of the first paragraph. For added emphasis, let it stand alone in its own paragraph. Then, in flash-back fashion, you can cover any details that will help readers to understand or act.

For example, the following engineering memo opens with a story and delays the point until the bitter end. Though the arrangement is not chaotic,

Start from how your engineering projects are

helping your company to improve the business

bottom line:

- improve quality, customer satisfaction, market

  share, profitability, and productivity,

- reduce cost, time to markets, cycle time, lead

  time, defects, and customer complaints.

FIGURE 6.1
Make your bottom line the top line.

neither is it efficient for readers. It forces them to wonder where the writing is going and what, if anything, is in it for them.

> On August 22, temperature measurement and display circuits were built and control software was written to use the added hardware making use of the HC11 analog-to-digital converter and the serial subsystems. In this design, the overall objectives were met. By keeping track of the measured temperature, the HC11 is able to control a temperature display that uses light emitting diodes. Also, if the temperature becomes very cold or hot, an alarm message is sent to a host PC terminal. This design has many potential applications, including temperature control and factory automation. This memo presents the design of a temperature measurement and display system that uses the Motorola HC11 micro-controller

The following revised version gets to the point quickly and explains in flashback fashion:

> This memo presents the design of a temperature measurement and display system that uses the Motorola HC11 microcontroller. This design makes use of the HC11 analog-to-digital converter and the serial subsystems. On August 22, temperature measurement and display circuits were built and control software was written to use the added hardware. In this design, the overall objectives were met. By keeping track of the measured temperature, the HC11 is able to control a temperature display that uses light emitting diodes. Also, if the temperature becomes very cold or hot, an alarm message is sent to a host PC terminal. This design has many potential applications, including temperature control and factory automation.

6.2 Purpose Statement and Blueprints

Even in short messages, the information may be so varied that no particular sentence stands out as more important than all the others. In this case, alert your readers to what is ahead by using a purpose statement or blueprint.

6.2.1 Writing Effective Purpose Statements

A purpose statement is a summary sentence that announces generally what is ahead: "Here is some background to help you prepare for Thursday's meeting with the product development team."

A purpose statement is a declarative sentence that summarizes the specific topic and goals of a document. It is typically included in the introduction to give the reader an accurate, concrete understanding of what the document will cover and what he or she can gain from reading it. To be effective, a statement of purpose should be:

- Specific and precise—not general, broad, or obscure
- Concise—one or two sentences
- Clear—not vague, ambiguous, or confusing
- Goal-oriented—stated in terms of desired outcomes

Some common introductory phrases for purpose statements include:

1. "The purpose of this paper/letter/document is to ... "
2. "In this paper, I will describe/explain/review the ... "
3. "My reason for writing is to ... "
4. "This paper will discuss the ... "
5. "The purpose of this paper is twofold: to ____ and ____."

Examples of ineffective purpose statements:

- "The purpose of this report is to describe the changes that are occurring in engineering design."
- Critique: too vague and broad. No clear expectation of what the reader will learn.
- Questions: What specific changes in engineering will be described? What types of changes? What aspects of engineering will be discussed? Will this report also discuss the effects of these changes?
- "The purpose of this report is to discuss the semiconductor failures."
- Critique: too vague and broad. It is not clear what aspect of the semiconductor failures will be discussed or what the reader will learn.

- Questions: What specific aspects of these semiconductor failures will be discussed? The causes of these failures? The signs or symptoms of these failures? The effects of these failures? If so, what types of effects—electrical, mechanical, functional?
- "This memo will cover the different ways a project can become organized."
- Critique: obscure and misleading. It is not clear what is meant by "different ways" or "become organized." These terms are vaguely stated and ambiguous.
- Questions: What is meant by "different ways" and "become organized"? What, specifically, will the reader learn about projects and how they become organized? Any specific types of organizations? Any specific types of projects?

Examples of effective purpose statements:

- "This memo will describe four common causes of embedded software failures in global positioning systems and explain how to use a five-step procedure to effectively prevent the failures."
- Critique: Very specific about what aspects of embedded software failures will be discussed. Very precise about how much information will be given. Very clear about what the reader will learn.
- "This report will explain how project leaders can use four planning skills to improve engineering productivity in new product development."
- Critique: Very specific about what will be discussed (planning strategies) and what the outcome will be for the reader (how to improve employee productivity).
- "The purpose of this report is to describe the main causes of traffic congestion in Chicago and required transportation network management."
- Critique: Leaves no doubt about the report's main purpose. Specific about the focus of the traffic congestion (Chicago).

6.2.2   Blueprinting: Planning Your Writing

A blueprint takes a purpose statement further by listing the points to be developed: "To help you prepare for Thursday's meeting with the product development team, here are some ways to field questions about deadlines and costs." Just as the blueprint of a building is the specific plan that will be used to guide construction efforts, the blueprint of a written engineering document is a tool that an engineer uses to define structure (see Figure 6.2).

It is a lot easier to add a new door, move a room from one side of the house to the other, or replace a whole story if you do it on the blueprint. When engi-

> Just as the blueprint of a building is the specific
>
> plan that will be used to guide construction efforts,
>
> the blueprint of an essay is a tool that an author
>
> uses in order to define structure. It's a lot easier to
>
> add a new door, move a room from one side of the
>
> house to the other, or replace a whole story if you
>
> do it on the blueprint.

FIGURE 6.2
Blueprint for writing.

neers create a specific plan for their project report before they start churning out paragraphs, they save themselves time that they might otherwise have to spend frantically crossing out and rewriting just before the due date.

The blueprint is a brief list of the points you plan to make, compressed into just a few words each, in the same order in which they appear in the body of your project report. As you introduce each new point, remind the reader of your key points, but avoid lengthy repetitions. Here are two tips for developing a blueprint:

- Develop your ideas in the order the blueprint uses.
- Repeat the blueprint's key points.

6.3 Open Long Reports with a Summary

Engineers are busy readers. An engineering document that runs three pages or more should open with a formal summary. Craft the summary carefully. You may write the summary last, but your audience needs to read it first. A summary matters most to the senior decision makers—those who may lack the time or interest to read further.

A summary is not a disguised background section. It gives the very best information from the rest of the document. Do not hold back. Do not say, "This report examines the need for new equipment evaluation forms," if you mean "We need a new equipment evaluation form."

A summary highlights the entire report in one, concise paragraph of about 100 to 200 words. It might be useful to think in terms of writing one sentence to summarize each of the traditional report divisions: objective, method, discussion, conclusions. Emphasize the objective, which states the problem, and the analysis of the results, including recommendations. Avoid the temptation to copy an entire paragraph from elsewhere in your report and make it do double duty. Because the summary condenses and emphasizes the most important elements of the entire report, you cannot write it until after you have completed the report. However, you should present it at the beginning of your report.

> Example: This engineering report for a U.S. government agency compares nuclear plants, fossil fuels, and solar generators, in order to determine which energy source will best meet the nation's needs. The criteria for comparison were the economic, social, and environmental effects of each alternative. The report concludes that nuclear energy is the best of these options, because North America is not self-sufficient in fossil fuels, and solar power is currently too unreliable for industrial use. Although nuclear plants need comprehensive risk management, nuclear energy is still the best short-term solution before other more advanced renewable energy becomes economically feasible.

> (Summary–Specific & Detailed)

Remember, the summary should be precise and specific; give details. A technical document is not a mystery novel; give your conclusion immediately, and support it later. You can do unusual things with summaries. To highlight a summary, print it on colored paper. To save time, send just the summary to those who need to see the document but not take action on it.

6.4 Use More Topic Sentences

A topic sentence states the main point of a paragraph: It serves as a mini thesis for the paragraph. You might think of it as a signpost or a headline for your readers—something that alerts them to the most important, interpretive points in your project report. When read in sequence, your report's topic sentences will provide a sketch of the idea or argument presented in the report. Thus, topic sentences help protect your readers from confusion by guiding them through the idea or argument. In addition, topic sentences can also help you to improve your project report by making it easier for you to recognize gaps or weaknesses in your idea or argument. A topic sentence performs at least some of the following functions:

- Announces the topic: "Forensic engineering investigates causes of failure."

- Makes a transition from the previous paragraph: "Similar to engineering requirements, ideas for engineering inventions are neither true nor false."
- Asks and answers a rhetorical question: "What about motion that is too slow to be seen by the human eye? That problem has been solved by the use of the time-lapse camera."—James C. Rettie
- Anticipates subtopics to be discussed in the paragraph or in an entire section of paragraphs: "Methods used in forensic engineering include reverse engineering, inspection of witness statements, solicitation of expert opinions, a working knowledge of current standards, as well as examination of the failed component itself."

A topic sentence usually appears at the very beginning of each paragraph. A well-organized paragraph supports or develops a single controlling idea, which is expressed in the topic sentence. A topic sentence has several important functions: It substantiates or supports an engineering writing's purpose statement; it unifies the content of a paragraph and directs the order of the sentences; and it advises the reader of the subject to be discussed and how the paragraph will discuss it. Readers generally look to the first few sentences in a paragraph to determine the subject and perspective of the paragraph. That is why it is often best to put the topic sentence at the very beginning of the paragraph. In some cases, however, it is more effective to place another sentence before the topic sentence—for example, a sentence linking the current paragraph to the previous one or a sentence to provide background information. Ask yourself what is going on in your paragraph.

- Why have you chosen to include certain information?
- Why is the paragraph important in the context of your argument?
- What point are you trying to make?

Although most paragraphs should have a topic sentence, a few situations arise when a paragraph might not need a topic sentence. For example, you might be able to omit a topic sentence if a paragraph narrates a series of events, if a paragraph continues developing an idea that you introduced (with a topic sentence) in the previous paragraph, or if all the sentences and details in a paragraph clearly refer—perhaps indirectly—to a main point. The vast majority of your paragraphs, however, should have a topic sentence.

6.5 Develop Headings

Use precise, informative headings to divide documents into sections. Headings are the titles and subtitles within the actual text of much professional scientific, technical, and business writing. They are similar to the parts of an outline that have been pasted into the actual pages of a report or other document. Headings are an important feature of professional technical writing: They alert readers to upcoming topics and subtopics, help readers find their way around in long reports and skip what they are not interested in, and break up long stretches of straight text.

When topics vary widely, use headings. Headings are useful for you in any kind of engineering writing, from manuals and reports to letters and e-mail. They keep you organized and focused on the topic. When you begin using headings, your impulse may be to insert the headings after you have written the rough draft. Instead, visualize the headings before you start the rough draft, and insert them as you write. Design headings so that their levels are clearly indicated. Use type size, type style, color, bold, italics, and alignment in such a way that the level of the heading is obvious. ("Levels" of headings are similar to levels in an outline: first level would correspond to the roman numerals; second level, to the capital letters; and so on.) Pay attention to the following:

- Make headings informative about sections they introduce. Pack each heading with lots of information. Headings such as "Technical Background" do not describe anything. How often have you seen several different sections titled "Applications" or "Appeals" in one set of documents? Applications for what? Appeals of what? If you say "Applications for Underground Mining Permits on Public Land," the reader knows exactly what you are talking about, and knows the difference between that section and another section titled "Applications for a Temporary Use Permit to Transport Cattle Across Public Land."
- Informative headings often start with such words as "when to" and "what to do if." For example,
    - "When to Establish a Failure Mode and Effect Analysis"
    - "What to Do If Getting a High-Risk Priority Number"
- Make headings parallel in phrasing. Parallelism sends readers important clues as to whether the section is similar in nature to the preceding ones.
    - "Why perform a reliability test?"
    - "How to request a reliability test"
- Avoid "lone headings"—it is the same concept as having an "A" without a "B" or a "1" without a "2" in outlines.
- Avoid "stacked headings"—this is two or more consecutive headings without intervening text.
- Avoid referring to headings with pronouns in the text that follows. If you have a heading such as "Configuring the Software," do not follow it with a sentence such as "This next phase ... ."

- Consider using the "hanging-head" format to make headings stand out more and to reduce the length of regular text lines. In the hanging-head design, some or all of the headings are on the left margin, while all text is indented one to two inches.

- Consider using "run-in" headings for your lowest-level heading. It can be difficult to rely solely on type style and size to indicate heading levels. A run-in heading "runs into" the beginning of a paragraph and ends with a period. You can use some combination of bold, italic, or color for these headings.

As with other headings, subject lines for memos or e-mail should be informative. Go beyond the general topic to your specific purpose. "Parking" might become "Request for Parking Spaces."

6.6   Structure Vertical Lists

Look for opportunities to divide paragraphs into vertical lists. Vertical lists highlight a series of requirements or other information in a visually clear way. Use vertical lists to help your reader focus on important material. Vertical lists:

- Highlight levels of importance
- Help the reader understand the order in which things happen
- Make it easy for the reader to identify all necessary steps in a process
- Add blank space for easy reading

Vertical lists are useful because they emphasize certain information in regular text. When you see a vertical list of three or four items strung out on the page, instead of in a normal paragraph format, you naturally notice it more and are likely to pay more attention to it. Certain types of lists also make for easier reading. For example, in inspection instructions for an assembly process, it is a big help for each step to be numbered and separated from the preceding or following steps. Vertical lists also create more white space and spread out the text so that pages do not seem like solid walls of words.

Vertical lists are also helpful in clarifying the chronological order of steps in a process.

When an engineer presents a completed Form for Test Request:

(a) Enter the engineer's department number.
(b) Fill out the test purpose, setup, and procedure.
(c) Keep a copy for the engineer's record.
(d) Send a copy to the test laboratory.

Yet, you can overuse vertical lists. Remember to use them to highlight important information, not to overemphasize trivial matters.

If you use bullets, use solid round or square ones. Bullets are not the place to be overly creative. Large, creative bullets with strange shapes tend to distract the reader. However, in some circumstances, numbers are a helpful option. You can best use numbers when you are highlighting the order of steps in a process, or when you are making a point that a certain number of items are included in the list. See the examples in Table 6.1.

Your lists will be easier to read if you:

- Always use a lead-in sentence to explain your lists.
- Indent your list from the lead-in sentence margin.
- Use left justification only—never center justification.

Table 6.2 illustrates that indented lists are easier to read than centered lists, and Table 6.3 shows that vertical lists are much more appealing visually and easier to read than running text. They make your documents appear less dense and make it easier to spot main ideas. They are also an ideal way to present items, conditions, and exceptions.

Similar to headings, the various types of vertical lists are an important feature of engineering writing: They help readers understand, remember, and review key points; they help readers follow a sequence of actions or events; and they break up long stretches of straight text. In engineering writing, you must use a specific style of lists, such the one discussed next:

- Use a list to highlight or emphasize text or to enumerate or make for easier reference.
- Use exactly the spacing, indentation, punctuation, and caps style shown in the discussion and illustrations that follow in the text.
- Make list items parallel in phrasing.
- Make sure that each item in the list agrees grammatically with the lead-in.
- Use a lead-in to introduce the list items, which indicates the meaning or purpose of the list, and punctuate it with a colon.
- Never use headings as lead-ins for lists.
- Avoid overusing lists; using too many lists destroys their effectiveness.
- Use similar types of lists consistently in similar text in the same document.
- Guideline for specific types of lists: Look for opportunities to divide paragraphs into vertical lists. You have the makings of a vertical list whenever three or more ideas share the same relationship with a larger idea.

TABLE 6.1

Using Numbers in a List

Chronological Order	Number of Items
Several steps should be applied: 1. Complete the test request. 2. Sign the test request. 3. Have your manager review and sign the test request. 4. Keep the bottom copy. 5. Return the rest of the test request to the engineering laboratory.	We need to know your opinion about the location of the engineering committee meeting. We narrowed it down to three choices: 1. Engineering Conference Room 2. Training Room 3. Engineering War Room
Numbering this list emphasizes that the engineer should keep a copy signed by the manager.	Numbering this list reinforces the three choices mentioned in the introductory sentence.

TABLE 6.2

Indenting vs. Centering

Centering—Harder to Read	Indenting—Easier to Read
Deliverables for design review • Design specification • Design failure mode and effect analysis • Design verification plan	Deliverables for design review When you come to the product design review, you should bring the following: • Design specification • Design failure mode and effect analysis • Design verification plan
Without a lead-in sentence, it is not clear who is to bring the deliverables for design review. Centering the bullets may make a nice pattern, but it makes it very difficult to see where statements begin and end.	In this sample, the lead-in sentence makes it clear who is to bring the deliverables to design review. Indenting makes it easier to see how the information is chunked.

TABLE 6.3

Vertical List vs. Running Text

Running Text—Harder to Read	Vertical List—Easier to Read
Each completed test request must contain a detailed statement including the following information: the test purpose, the test set-up, the equipment calibration, the test procedure, and the test data collection form.	With your test request, provide the following information: (a) Test purpose (b) Test set-up (c) Equipment calibration (d) Test procedure (e) Test data collection form

Bibliography

Alred, G. J., *Business Writer's Handbook*. 6th ed. New York: St. Martin's, 2000.

American Management Association, *The AMA Style Guide for Business Writing*. New York: AMACOM, 1996.

Beamer, L. and Varner, I. I., *Intercultural Communication in the Global Workplace*. 2nd ed. Boston: McGraw-Hill/Irwin, 2001.

Chaney, L. H. and Martin, J. S., *Intercultural Business Communication*. 2nd ed. Upper Saddle River, NJ: Prentice Hall, 2000.

Cialdini, R. B., *Influence: The Psychology of Persuasion*. Rev. ed. New York: Morrow, 1993.

Downey, R., Boland, S., and Walsh, P., *Communications Technology Guide for Business*. Boston: Artech House, 1998.

Eckhouse, B. E., *Competitive Communication: A Rhetoric for Modern Business*. Rev. ed. New York: Oxford University Press, 1999.

Fearn-Banks, K., *Crisis Communications: A Casebook Approach*. 2nd ed. Mahwah, NJ: Lawrence Erlbaum Associates, 2002.

Gardner, H., *Changing Minds: The Art and Science of Changing Our Own and Other People's Minds*. Boston: Harvard Business School Press, 2004.

Geffner, A. B., *How to Write Better Business Letters*. 3rd ed. Hauppauge, NY: Barron's, 2000.

Geffner, A. B. and NetLibrary, Inc., *Business English: A Complete Guide to Developing an Effective Business Writing Style*. 3rd ed. Hauppauge, NY: Barron's Educational Series, 1998.

Guilar, J. D., *The Interpersonal Communication Skills Workshop: Listening, Assertiveness, Conflict Resolution, Collaboration*. New York: AMACOM, 2001.

Harvard Business Essentials: Business Communication. Boston: Harvard Business School Press, 2003. (The Harvard Business Essentials series.)

Harvard Business Review on Effective Communication. Boston: Harvard Business School Press, 1999. (The Harvard Business Review paperback series.)

Holtz, S., *Corporate Conversations: A Guide to Crafting Effective and Appropriate Internal Communications*. New York: AMACOM, 2004.

Jablin, F. M. and Putnam, L., *The New Handbook of Organizational Communication: Advances in Theory, Research, and Methods*. Thousand Oaks, CA: Sage Publications, 2001.

Levine, R. V., *The Power of Persuasion: How We're Bought and Sold*. Hoboken, NJ: John Wiley & Sons, 2003.

McLeary, J. W., *By the Numbers: Using Facts and Figures to Get Your Projects and Plans Approved*. New York: American Management Association, 2000.

Munter, M., *Guide to Managerial Communication: Effective Business Writing and Speaking*. 4th ed. Upper Saddle River, NJ: Prentice Hall.

Pan, Y., Scollon, S. B. K., and Scollon, R., *Professional Communication in International Settings*. Malden, MA: Blackwell Publishers, 2002.

Pearce, T., *Leading out Loud: Inspiring Change through Authentic Communication*. New and rev. ed. San Francisco: Jossey-Bass Publishers, 2003.

Rankin, E., *The Work of Writing: Insights and Strategies for Academics and Professionals*. San Francisco: Jossey-Bass, 2001.

Ryan, K., *Write up the Corporate Ladder: Successful Writers Reveal the Techniques That Help You Write with Ease and Get Ahead*. New York: American Management Association, 2003.

Simmons, J., *We, Me, Them, & It: The Powers of Words in Business*. New York, and London: Texere, 2002.

Stockard, O., *The Write Approach: Techniques for Effective Business Writing*. San Diego: Academic Press, 1999.

Tingley, J. C., *The Power of Indirect Influence*. New York: AMACOM, 2001.

Whalen, D. J., *I See What You Mean: Persuasive Business Communication*. Thousand Oaks, CA: Sage Publications, 1996.

Wiener, V., *Power Communications: Positioning Yourself for High Visibility*. New York: New York University Press, 1994.

Weissman, J., *Presenting to Win: Persuade Your Audience Every Time*. Upper Saddle River, NJ: Financial Times/Prentice Hall, 2003.

Worth, R., *Webster's New World Business Writing Handbook*. Indianapolis: Wiley Publishing, 2002.

7

Write As If Talking to Your Engineering Associates

The most readable engineering writing sounds like engineers talking to engineers, because engineering professionals "hear" writing more efficiently just reading it. Research indicates that a conversational writing style is generally more effective at producing learning results than more formal writing. Make your writing more like speaking (see Figure 7.1). If you want your engineering readers to learn and remember what you write, say it conversationally. This refers to engineering memos, requests, procedures, instructions, and reports. Messages written in a conversational style are more likely to be retained and recalled than a report on the same topics written in a more formal tone. Most of us know this intuitively, but several studies have proven it.

A conversational style is the most effective form of engineering writing. Why is it so effective? The answer is simple. Your busy engineering reader does not have to struggle to understand the message. This "brain comfort" is important when you consider all the different messages that are vying for our attention every day. To achieve a spoken style, imagine that your engineering readers are sitting across from you. Then use the techniques discussed in the following sections—the best of speaking. Once you have written the message, read it aloud. It should sound like something you might hear in person. Think of writing not only as words on a page, but also as a talk with your engineering readers.

> Imagine your readers are sitting across from you.
>
> You speak to them with personal pronouns,
>
> everyday words, transitions, and short
>
> sentences. Ask them questions and help them
>
> answer the questions in a good tone and concrete
>
> "spoken" language.

FIGURE 7.1
Writing as if speaking.

7.1   Use Personal Pronouns

Rudolph Flesch, a pioneering advocate of readability, put great stock in the liveliness of the written word. One way of getting that liveliness into our writing, he said, is to use the personal pronouns: you, me, I, we, us, he, she, him, her, and they.

Most of us know that personal pronouns should be used in engineering requests and memos. Research indicates that they increase readability in all kinds of engineering writing, such as proposals, test instructions, and even technical reports. Follow these principles:

1. Use we, us, and our when speaking for your project team.
2. Use I, me, and my when speaking for yourself.
3. Use you, stated or implied, to refer to your engineering readers.

In choosing among pronouns, show more interest in your readers than yourself by favoring "you" in engineering writings. "You" reinforces the message that the document is intended for your reader in a way that "he," "she," or "they" cannot. More than any other single technique, using "you" pulls readers into your document and makes it relevant to them. Using "we" to refer to your agency makes your sentences shorter and your document more accessible to readers.

A preference for "you" increases the chances that you will see things from your engineering readers' viewpoints. When we use personal pronouns, several important things happen. For starters, we personalize our writing, and that makes it easier for readers to relate to the subject. For example, which of the following two sentences would be more effective?

1. The use of engineering report templates standardizes written communication format.
2. When we use engineering reporting templates, we standardize our writing format.

I think you would agree that sentence 2, with several personal pronouns, is livelier and more likely to be understood. That sentence allows readers and listeners to relate to the words; in other words, a personal connection is established. On the other hand, the first sentence is a collection of abstract concepts.

In using personal pronouns, we also make our writing more like our spoken communication. Listen to almost any conversation and you will notice frequent use of "I," "You," and "We." It is quite natural to speak that way. When you write to a team of engineers but no one in particular, imagine you are talking to the group or, even better, an individual team member. Only one engineer reads your writing at any given time; therefore, the most readable engineering writing speaks directly to an individual. The key is to use "you" and "your," whether stated or implied.

For

All who plan to use Genesis Test Chamber in September are required to complete the enclosed test request.

Try

If you plan to use Genesis Test Chamber in September, please complete the enclosed test request.

Try personal pronouns yourself. Take an engineering document that you want others to read and rewrite it to include more of them. In the process of doing this, you are bound to make it more readable. In addition, you will make your writing more effective.

7.2 Rely on Everyday Words

The complexity of the engineering project and the need for precision processes require some complex words. However, do not overuse highly technical or scientific language in place of equally appropriate everyday words.

How a word is finally understood depends on many factors—not all of which are under your control. When writing an engineering document, you must anticipate how your readers will respond to the word, anticipating whether certain words are within their vocabularies. Failure to take the reader sufficiently into account often shows up in a writer's abuse of jargon, which, in one sense, means the specialized vocabulary of people in a particular group or profession and, in a broader sense, means the use of highly technical or scientific language in place of equally appropriate everyday words.

Keep your writing style simple. Follow the standard rule in all editing: Prefer the simpler word to the complex word. This means using the words "extra" or "more" instead of "additional;" "help" instead of "assistance;" and "use" instead of "utilize." Although you might need specialized or technical words, depending on your subject, you should choose the simpler word instead of the more difficult word whenever you can.

Your engineering writing should be accurate yet easy to read. In addition to the advice about preferring the simpler word to the complex word, you can also make your writing more readable if you reduce "heavy" words. The traditional advice is to reduce the number of three-syllable and four-syllable words.

For

This demonstrates an understanding of the development of the new helicopter, the need for global engineering teams, and a multicultural perspective.

Try

This shows we understand how to design and build the new helicopters, and know why we need to work together and respect our different cultural backgrounds.

Breaking the habit of using heavy-sounding words and replacing them with shorter, everyday words helps make your writing easier to read and encourages you to explain your ideas in more specific terms. Your choice of words should be based on what will be clearer for your reader. If you are not sure, ask. Test out your document with some of the people who are likely to use it. To help you draft easy-to-understand documents, here are some guidelines for your choice of words.

Use simple, familiar words instead of unfamiliar words. Write as if someone is asking you what you mean. Here are a few examples of simple words and phrases you might substitute:

For	Try
assistance	help
accomplish	do
ascertain	find out
disseminate	send out, distribute
endeavor	try
expedite	hasten, speed up
facilitate	make easier, help
formulate	work out, devise, form
in lieu of	instead of
locality	place
optimum	best, greatest, most
strategize	plan
utilize	use

7.3   Use Short, Spoken Transitions

Every time you introduce a new main section or topic, you need a transition. Transitions are signals or cues in your writing that show your readers the relationship between one item and the next. Transitions help the reader see that a certain train of thought is being continued, developed, challenged, changed, or summarized. Transitions make your writing flow and make it easier for your reader to understand your ideas. Use a transition to link paragraphs by putting a transition word or phrase in the first sentence of the new paragraph, or to link sentences within a paragraph.

7.3.1 The Function and Importance of Transitions

In engineering writing, your goal is to convey information clearly and concisely, if not to convert the reader to your way of thinking. Transitions help you to achieve these goals by establishing logical connections between sentences, paragraphs, and sections of your writings. In other words, transitions tell readers what to do with the information you present them. Whether single words, quick phrases, or full sentences, they function as signs for readers that tell them how to think about, organize, and react to old and new ideas as they read through what you have written.

Transitions signal relationships between ideas such as: "Another example coming up—stay alert!" or "Here's an exception to my previous statement" or "Although this idea appears to be true, here's the real story." Basically, transitions provide the reader with directions for how to piece together your ideas into a logically coherent argument. Transitions are not just "window dressing" that decorates your writing by making it sound or read better. They are words with particular meanings that tell the reader to think and react in a particular way to your ideas. In providing the reader with these important cues, transitions help readers understand the logic of how your ideas fit together.

7.3.2 How Transitions Work

The organization of your writing includes two elements: (1) the order in which you have chosen to present the different parts of your discussion or argument, and (2) the relationships you construct between these parts. Transitions cannot substitute for good organization, but they can make this organization clearer and easier to follow.

Transitions can help reinforce the underlying logic of your writing's organization by providing the reader with essential information regarding the relationship between your ideas. In this way, transitions act as the glue that binds the components of your argument or discussion into a unified, coherent, and persuasive whole. Effectively constructing each transition often depends upon your ability to identify words or phrases that will indicate for the reader the kind of logical relationships you want to convey. Opt for short, spoken transitions. Short transitions help set an ordinary tone for your writing. Here is a sample list of some alternative words for common, wordy transitions:

For	Try
consequently	so
with regard to	about
by means of	by
in the event that	if
until such time	until
during such time	while
in respect of	for
in view of the fact	because
on the part of	by
subsequent to	after
under the provisions of	under
with a view to	to
it would appear that	apparently
it is probable that	probably
notwithstanding the fact that	although
adequate number of	enough
excessive number of	too many

7.4   Keep Sentences Short

Effective writing builds ideas from sentence to sentence. The simple, declarative sentence is the easiest way to process information. Sentences that differ from that simple structure may cause readability problems.

Readers can only take in so much new information at one time. Express only one idea in each sentence. Long, complicated sentences often mean that you are not clear about what you want to say. Shorter sentences show clear thinking. Shorter sentences are also better for conveying complex information; they break the information up into smaller, easier-to-process units. In technical documents, keep your average sentence between 10 and 20 words. You may go as low as 10 or 11 words if you are writing instructions with many short, concise sentences that tell the user what to do.

For

This process does not appear to be well understood by program management, even though this group of people has primary responsibility for implementing the process (*25 words*).

Try

The program managers who are most responsible for performing this process do not seem to understand it well (*19 words*).

Readers can understand longer sentences if they are well constructed and use familiar terms. For variety, mix long sentences and short ones. A variety of sentence lengths makes your writing more interesting. Vary your sentence structure to avoid choppiness, but do not revert to tangled multi-clause sentences.

7.5 Reach Out to Your Engineering Readers by Asking Questions

Look for opportunities to reach out to your engineering readers by asking questions. Similar to short, concise sentences, questions narrow the distances between you and your engineering readers. As often as possible, write section headings as questions. Try to ask the questions your readers would ask. Answer each question immediately.

When your engineering associates read a question, they feel you are talking to them. Engineering requests are easy to phrase as questions, and their focus invites concise answers.

For

Your acknowledging receipt of the new test equipment would be appreciated.

Try

Has the new test equipment arrived? Please let us know.

Using the Question and Answer (Q&A) format helps readers to scan the document and find the information they want. It also increases the chances that they will see a question that they did not realize they had, but to which they need to know the answer. This format will be enormously helpful to your engineering readers.

7.6 "5 Whys"—A Technique for Engineering Problem Solving

"If you don't ask the right questions, you don't get the right answers. A question asked in the right way often points to its own answer. Asking questions is the ABC of diagnosis. Only the inquiring mind solves problems."

Edward Hodnett

Asking "Why?" may be a favorite technique of your three-year-old child, and, although it can drive you crazy, it could teach you a valuable tool for engineering problem solving. The 5 Whys is a technique used in analyzing engineering problems.

7.6.1   What Is "5 Whys"?

The 5 Whys is a simple problem-solving technique that helps users to get to the root of a problem quickly. Made popular in the 1970s by the Toyota Production System, the 5 Whys strategy involves looking at any problem and asking: "Why?" and "What caused this problem?"

Very often, the answer to the first "why" will prompt another "why;" the answer to the second "why" will prompt another; and so on, thus the name "5 Whys."

The 5 Whys typically refers to the practice of asking, five times, why an engineering problem has occurred in order to get to the root cause or causes of the problem. There can be more than one cause to a problem as well. In an organizational context, root cause analysis is generally performed by a project team for problem solving. No special technique is required.

By repeatedly asking the question "Why" (five is a good rule of thumb), you can peel away the layers of symptoms that can lead to the root cause of a problem. Very often, the ostensible reason for a problem will lead you to another question. Although this technique is called "5 Whys," you may find that you will need to ask the question fewer or more times than five before you find the issue related to a problem.

7.6.2   What Are the Benefits of the 5 Whys?

It attempts to analyze a problem or issue by asking a series of "Why (did this happen)?" questions. The benefits of 5 Whys include the following:

- Help identify the root cause of a problem quickly
- Determine the relationship between different root causes of a problem
- One of the simplest tools
- Easy to learn and apply
- Easy to complete without mathematical statistical analysis

7.6.3   When Is 5 Whys Most Useful?

5 Whys is a simple tool that addresses single-problem events instead of broad program organizational issues:

- When problems involve human factors or interactions
- Commonly used in daily engineering work

- Can be used for developing new products and improving existing products

7.6.4 How to Complete the 5 Whys

If a problem occurs, the first "Why?" question is asked: "Why did this happen?" A number of answers may be found, and for each of these the next "Why?" is asked: "Why is that?" The whole process is repeated until five consecutive "Why?"s have been asked and answered. In most instances, it has been found that five repeated whys are necessary to get to the real root cause of the problem.

1. Write down the specific problem. Writing the issue helps you formalize the problem and describe it completely. It also helps a team focus on the same problem.
2. Ask why the problem happens, and write the answer below the problem.
3. If the answer you just provided does not identify the root cause of the problem that you wrote in Step 1, ask why again and write down that answer.
4. Loop back to Step 3 until the team is in agreement that the problem's root cause is identified. Again, this may take fewer or more attempts than 5 Whys.

7.6.5 5 Whys Examples

As illustrated by the following examples, the 5 Whys method is to ask why an event happened and place the resulting answer in the cause chain. Each question should be simple, short, and focused on a single question, and should begin with "Why ... ?" After each cause has been identified, the next question to ask is "Why did this event happen?"

> **Problem Statement: A scanning electron microscope (SEM) was found in use, beyond its calibration date, on the shop floor.**

1. Why was an SEM in use beyond its calibration date?
   Because the SEM was not recalled and the technician did not check the calibration label.
2. Why was the SEM not recalled?
   Because the SEM was not on the recall list.
3. Why was the SEM not on the recall list?
   Because the SEM was just recently purchased.
4. Why are new SEMs not added to recall list?
   Because no procedure or specific training on purchasing gauges has been established.

In this case, only four Whys were required to find out that a non-value-added signature authority is helping to cause a process breakdown. Let us now take a look at a slightly more humorous example.

Problem Statement: You are on your way home from work, and your car stops in the middle of the road.

1. Why did your car stop?
   Because it ran out of gas.
2. Why did it run out of gas?
   Because I didn't buy gas on my way to work.
3. Why didn't you buy gas this morning?
   Because I didn't have any money.
4. Why didn't you have any money?
   Because I left my wallet at home.
5. Why did you leave your wallet at home?
   Because I was running late.

As you can see, in both examples, the final Why leads the team to a statement (root cause) upon which the team can take action. It is much quicker to develop a system that keeps the sales director updated on recent sales or teach a person to "bluff" a hand than it is to try to directly solve the preceding problems without further investigation.

The five iterations are not gospel; instead, it is postulated that five iterations of asking why is generally sufficient to get to a root cause. The real key is to encourage the troubleshooter to avoid assumptions and logic traps and instead to trace the chain of causality in direct increments from the effect through any layers of abstraction to the first or root cause.

The 5 Whys technique was originally developed by Sakichi Toyoda and was later used within Toyota Motor Corporation during the evolution of their manufacturing methodologies. It is a critical component of problem-solving training delivered as part of the induction into the Toyota Production System. The architect of the Toyota Production System, Taiichi Ohno, described the 5 Whys method as " … the basis of Toyota's scientific approach … by repeating why five times, the nature of the problem as well as its solution becomes clear." The tool has seen widespread use beyond Toyota, and is now used by engineering professionals of all disciplines.

Bibliography

Alred, G. J., *Business Writer's Handbook*. 6th ed. New York: St. Martin's, 2000.

American Management Association, *The AMA Style Guide for Business Writing*. New York: AMACOM, 1996.

Beamer, L. and Varner, I. I., *Intercultural Communication in the Global Workplace.* 2nd ed. Boston: McGraw-Hill/Irwin, 2001.

Chaney, L. H. and Martin, J. S., *Intercultural Business Communication.* 2nd ed. Upper Saddle River, NJ: Prentice Hall, 2000.

Cialdini, R. B., *Influence: The Psychology of Persuasion.* Rev. ed. New York: Morrow, 1993.

Downey, R., Boland, S., and Walsh, P., *Communications Technology Guide for Business.* Boston: Artech House, 1998.

Eckhouse, B. E., *Competitive Communication: A Rhetoric for Modern Business.* Rev. ed. New York: Oxford University Press, 1999.

Fearn-Banks, K., *Crisis Communications: A Casebook Approach.* 2nd ed. Mahwah, NJ: Lawrence Erlbaum Associates, 2002.

Gardner, H., *Changing Minds: The Art and Science of Changing Our Own and Other People's Minds.* Boston: Harvard Business School Press, 2004.

Geffner, A. B., *How to Write Better Business Letters.* 3rd ed. Hauppauge, NY: Barron's, 2000.

Geffner, A. B. and NetLibrary, Inc., *Business English: A Complete Guide to Developing an Effective Business Writing Style.* 3rd ed. Hauppauge, NY: Barron's Educational Series, 1998.

Guilar, J. D., *The Interpersonal Communication Skills Workshop: Listening, Assertiveness, Conflict Resolution, Collaboration.* New York: AMACOM, 2001.

Harvard Business Essentials: Business Communication. Boston: Harvard Business School Press, 2003. (The Harvard Business Essentials series.)

Harvard Business Review on Effective Communication. Boston: Harvard Business School Press, 1999. (The Harvard Business Review paperback series.)

Holtz, S., *Corporate Conversations: A Guide to Crafting Effective and Appropriate Internal Communications.* New York: AMACOM, 2004.

Jablin, F. M. and Putnam, L., *The New Handbook of Organizational Communication: Advances in Theory, Research, and Methods.* Thousand Oaks, CA: Sage Publications, 2001.

Levine, R. V., *The Power of Persuasion: How We're Bought and Sold.* Hoboken. NJ: John Wiley & Sons, 2003.

McLeary, J. W., *By the Numbers: Using Facts and Figures to Get Your Projects and Plans Approved.* New York: American Management Association, 2000.

Munter, M., *Guide to Managerial Communication: Effective Business Writing and Speaking.* 4th ed. Upper Saddle River, NJ: Prentice Hall.

Ohno, T., *Toyota Production System: Beyond Large-Scale Production.* Portland, OR: Productivity Press, 1988.

Pan, Y., Scollon, S. B. K., and Scollon, R., *Professional Communication in International Settings.* Malden, MA: Blackwell Publishers, 2002.

Pearce, T., *Leading out Loud: Inspiring Change through Authentic Communication.* New and rev. ed. San Francisco: Jossey-Bass Publishers, 2003.

Rankin, E., *The Work of Writing: Insights and Strategies for Academics and Professionals.* San Francisco: Jossey-Bass, 2001.

Ryan, K., *Write up the Corporate Ladder: Successful Writers Reveal the Techniques That Help You Write with Ease and Get Ahead.* New York: American Management Association, 2003.

Simmons, J., *We, Me, Them, & It: The Powers of Words in Business.* New York and London: Texere, 2002.

Stockard, O., *The Write Approach: Techniques for Effective Business Writing.* San Diego: Academic Press, 1999.

Tingley, J. C., *The Power of Indirect Influence*. New York: AMACOM, 2001.
Whalen, D. J., *I See What You Mean: Persuasive Business Communication*. Thousand
    Oaks, CA: Sage Publications, 1996.
Wiener, V., *Power Communications: Positioning Yourself for High Visibility*. New York:
    New York University Press, 1994.
Weissman, J., *Presenting to Win: Persuade Your Audience Every Time*. Upper Saddle
    River, NJ: Financial Times/Prentice Hall, 2003.
Worth, R., *Webster's New World Business Writing Handbook*. Indianapolis: Wiley
    Publishing, 2002.

8

"Trim" Your Expressions

8.1 Introduction

An effective written engineering document is similar to an efficient manufacturing facility—lean. Too much fat in our writing is not good for our engineering readers. Most engineers are busy and impatient people. They hate
to wait.

Engineering writing should be concise. Just as an engineering drawing
should have no unnecessary lines and a machine no unnecessary parts, a
sentence should contain no unnecessary words and a paragraph no unnecessary sentences. "Trimming" your expression communicates the engineering
issues or problems more clearly, because problems become visible instead of
hiding in the wordy expression (see Figure 8.1).

Engineering writing should be accurate and to the point. If you learn to
write concisely, your reports will be clearer and more readable. As presented
in Table 8.1, wordiness creates distractions. It is unnecessarily time-consuming
for the reader without adding any benefit.

Engineering writing should be concise.

- A sentence contains no unnecessary

  words.

- A paragraph contains no unnecessary

  sentences.

- A drawing contains no unnecessary

  lines.

FIGURE 8.1
Lean Expression: focusing on problem solving.

TABLE 8.1

Wordiness Creates Distractions

Wordy	Concise
Because there are many serious accidents on our roads, there are Road Risk Engineering and Management Programs. The aim of these programs is to cut down on the number of accidents on the roads, and to reduce the number of serious accidents.	Road Risk Engineering and Management Programs aim to reduce the frequency and severity of accidents.

Some engineers mistakenly think that wordy writing gives their work a more formal tone; thus, they "pad" their sentences with unnecessary verbiage. Remember that clear, concise expression of your ideas is preferred because wordiness, in addition to being distracting, can be annoying and insulting to your reader. Here are six practical rules to trim your expressions:

1. Prefer familiar words to the strange.

2. Prefer concrete word to the abstract.

3. Prefer the single word to the redundant.

4. Prefer the short word to the long.

5. Prefer the specific word to the ambiguous.

6. Prefer the strong word to the weak.

8.2   Prune Wordy Expressions

Avoid saying the same thing twice. Phrases that repeat, such as "true fact," "twelve noon," "I saw it with my own eyes," are wordy expressions. Wordy phrases are bad habits just waiting to take control of your business writing. Examples of wordy expressions are given in Table 8.2.

Using too many words often makes it difficult to understand what is being said. Wordy expressions force your engineering reader to work hard to figure out what is happening. In many cases, they may simply decide it is not worth the effort.

If you can take away words from your sentence and have it say the exact same thing, then it needs pruning. Replace wordy expressions with shorter substitutes. Avoid the wordy phrases in the Table 8.3. Use the shorter, simpler expressions in the right column.

TABLE 8.2

Examples of Wordy Expressions

Example of a Wordy Expression	Comment
In biology engineering, the molecules are linked together by van der Waals bonds.	If two things are linked, they must be together; thus, the use of the word "together" is unnecessary.
The past history of its technical use is not well documented.	Anything that is historical must have occurred in the past. In this sentence, "past" is unnecessary.
The reactor core was completely surrounded by water.	"Surrounded" means all around; the word "completely" is unnecessary.
Wearing radioactive protection in the engineering laboratory is absolutely essential when working with some isotopic species.	Superfluous words such as "absolutely," "virtually," and "actually" add nothing to the meaning of a sentence.

TABLE 8.3

Use the Lean Version

Wordy Expression	Lean Version
with regard to	about
by means of	by
in the event that	if
until such time	until
during such time	while
in respect of	for
in view of the fact	because
subsequent to	after
under the provisions of	under
with a view to	to
it would appear that	apparently
notwithstanding the fact that	although
adequate number of	enough
excessive number of	too many

8.3   Use Strong Verbs

Engineering is powerful. It moves the world ahead. Engineering writing should also be powerful. The most important word in a sentence is the verb, which should provide the power for effective actions.

Similar to engines, verbs are especially powerful because they convey action. They tell what is happening. A strong, simple, active verb attracts

TABLE 8.4

Turn a Nearby Word into a Specific Verb

Weak and Wordy	Strong and Precise
We held a meeting to give consideration to the proposal. (9 words)	We met to consider the proposal. (6 words).
The engineering managers made the decision to give their approval to the plan. (12 words)	The engineering managers decided to approve the plan. (8 words)
The team does research on new refrigeration technology. (8 words)	The team researches new refrigeration technology. (6 words)
This instrument will give the ability to perform semiconductor failure analysis. (11 words)	This instrument will enable us to analyze semiconductor failures. (9 words)
We need to perform a periodic review of the circuit design. (11 words)	We need to review the circuit design periodically. (8 words)

attention and resonates with engineering experience. Strong verbs say more in less time. You can eliminate a host of writing problems by using strong, precise verbs in your writing.

Weak verbs dilute your work. They generate wordiness, misplaced modifiers, and vagueness—none of which contribute to effective writing. Weak writing relies on general verbs, which take extra words to complete their meaning. Challenge the general verbs, such as "make" and "take." From Table 8.4, see whether you can turn a nearby word into a specific verb.

Forms of the verb "to be" yield the weakest verbs because they only denote that something exists. They do not fix meaning with any power and precision. And because the verb is weak, you usually have to add more words to sharpen the meaning of the sentence—often strings of prepositional phrases or modifiers. While you might wonder what the meaning of "to" is, it marks the most common verb we use in engineering writing.

Here are its eight forms: "be," "am," "is," "are," "being," "was," "were," "been." When you rely on them too heavily, you can clutter your prose with wordy, ineffective constructions. Normally, you might find a good verb hiding in the sentence as some other part of speech. See the examples in Table 8.5.

Look through your writing to find "to be" verbs. Can you change the sentence to clear away these verbs? Can you rearrange the sentence so that the verb is stronger?

8.4 Cut Doublings and Noun Strings

Redundant phrases are bad habits just waiting to take control of your writing. Doublings are repetitions that waste the reader's time. They add no or little

TABLE 8.5

Find a Good Verb for Strong and Precise Expression

Weak and Wordy	Strong and Precise
Oil is the dominant sector in Kuwait's economy. (8 words)	Oil dominates Kuwait's economy. (4 words)
The guidelines are applicable to engineers who make use of the new software program. (14 words)	The guidelines apply to engineers who use the new program. (10 words)
There is a lack of will by these team members to resolve the issues. (14 words)	These team members lack the will to resolve the issues. (10 words)
This project is made up of five segments. (8 words)	This project consists of five segments. (6 words)

TABLE 8.6

Cut Doublings

Wordy with Doublings	Concise
Your help and support led to important and significant process improvement.	Your help led to important process improvement.
We must comply with the standards and criteria for controlling and reducing Electromagnetic Interference (EMI).	We must comply with the standards for reducing Electromagnetic Interference (EMI).

TABLE 8.7

Avoid Long Clots of Nouns and Modifiers

Excessive Use of "Adjective–Noun Strings"	Add Some Words or Rewrite Entirely
Cost saving training needs planning survey.	Survey of training is needed for cost saving.
Commercial power systems safety reliability review meeting	Review meeting for safety and reliability of commercial power systems.

information. Avoid writing about a project's importance and significance when importance will do. Pairs of words with similar meanings add needless bulk to writing. Whatever the differences between synonyms, they are usually minor and not worth highlighting (see Table 8.6).

Though you should cut needless words, sometimes you can go too far. Avoid excessive use of "adjective–noun strings" (i.e., long strings of words that stack up in an attempt to modify a single word). They are hard to decipher— readers cannot tell what modifies what. Reading these strings, you may wonder how the pieces fit together or where they all will end. As presented in Table 8.7, you can prevent the problem by adding some words or rewriting entirely.

Avoid overusing noun forms of verbs. Use verbs when possible instead of noun forms known as nominalizations. Sentences with many nominalizations usually have forms of "be" as the main verbs. Using the action verbs

TABLE 8.8

Using Action Verbs to Prevent Dull Prose

Overused Noun Forms of Verbs	Use Verbs to Create Engaging Prose
The function of the department is the analysis of customer feedback.	This department analyzes customer feedback.
The current focus of the reliability engineering profession is failure prevention.	The reliability engineering profession currently focuses on failure prevention.

TABLE 8.9

Change Words Ending in "-ion" and "-ment" into Verb Forms

Use Words Ending in "-ion" and "-ment"	Change into Verb Forms
The settlement of product liability claims requires a test of control reliability.	Settling product liability claims requires testing control reliability.
Use that template for the preparation of the test report.	Use that format to prepare the report.

disguised in nominalizations as the main verbs, instead of forms of "be," can create engaging rather than dull prose (see Table 8.8).

Words ending in "-ion" and "-ment" are verbs turned into nouns. Whenever the context permits, change these nouns into verb forms. As presented in Table 8.9, these changes will shorten your sentences.

8.5   Eliminate Unnecessary Determiners and Modifiers

Engineers sometimes clog their writing with one or more extra words or phrases that appear to determine narrowly or to modify the meaning of a noun. These determiners and modifiers do not actually add to the meaning of the sentence. Although such words and phrases can be meaningful in the appropriate context, they are often used as "filler" and can easily be eliminated. Here are some examples:

The substance was boiled until it was reduced down to one third of its original volume.

The tension strips at the end are shorter in length and will not buckle, as will the longer compression strips near the center of the plate.

The glass construction above the old chimney in Melbourne Central is conical in shape.

Engineers today are often called upon and asked to advise on occupational health and safety issues.

The approval of the steering committee was absolutely essential before the modifications could be made to the wall.

TABLE 8.10

Convert Phrases into Single Words

Using Phrases—Wordy	Convert into Single Words—More Concise
The engineer with creativity...	The creative engineer....
The team showing the best performance...	The best-performing team...
Sonny Lee, our V.P. of engineering, suggested at our last board meeting the installation of microfilm equipment in the department of data processing.	At our last board meeting, VP Engineering Sonny Lee suggested that we install microfilm equipment in the data processing department.
As you carefully read what you have written to improve your wording and catch small errors of spelling, punctuation, and so on, the thing to do before you do anything else is to try to see where a series of words expressing action could replace the ideas found in nouns rather than verbs.	As you edit, first find nominalizations that you can replace with verb phrases.

The Sydney Monier aqueducts, which are still standing now today, were built in the late 1800s.

Each and every engineer should have good communication skills.

8.6   Change Phrases into Single Words

Using phrases to convey meaning that could be presented in a single word contributes to wordiness. Table 8.10 converts phrases into single words whenever possible.

8.7   Change Unnecessary Clauses into Phrases or Single Words

Be alert for clauses or phrases that can be pared to simpler, shorter constructions. Using a clause to convey meaning that could be presented in a phrase or even a word contributes to wordiness. The "that, who, and which clause" can often be shortened to a simple adjective. Table 8.11 converts modifying clauses into phrases or single words whenever possible.

8.8   Avoid Overusing "It is" and "There is"

Unless the "it" refers to something mentioned earlier, write around "it is." These expletive constructions delay meaning, lead to passive verbs, and hide

TABLE 8.11

Example: Convert Modifying Clauses into Phrases or Single Words

Use Modifying Clauses—Wordy	Convert into Phrases or Single Words—More Concise
The report, which was released recently ...	The recently released report ...
The system that is most efficient and accurate ...	The most efficient and accurate system ...

TABLE 8.12

Eliminating the Expletive Opening

Overuse "It is/was" and "There is/are"—Wordy	Eliminating the Expletive Opening—More Concise
It is the chief engineer who approves a system specification.	The chief engineer approves a system specification.
It is my understanding that your project covers new HVAC for residential buildings.	I understand that your project covers new HVAC for residential buildings.
There are four design rules that should be observed:	Four rules should be observed:
There are some project deadlines that cannot be changed.	Some deadlines cannot be changed.
It was recognized that there would be changes in priorities as the program develops.	We recognized that priorities would change as the program develops.

responsibility. Take the following example: "It is imperative that we find a solution." The same meaning could be expressed with this more succinct wording: "We must find a solution." Similar to "it is" constructions, forms of "there is" make sentences start slowly. Do not write these delayers without first trying to avoid them. You should generally avoid excessive or unnecessary use of expletives.

An expletive construction is a common device that often robs a sentence of energy before it gets a chance to do its work. The most common kind of unnecessary expletive construction involves an expletive followed by a noun and a relative clause beginning with that, which, or who. Table 8.12 demonstrates that you can create a more concise sentence by eliminating the expletive opening, making the noun the subject of the sentence, and eliminating the relative pronoun.

8.9 Eight Steps for Lean Writing

Engineering writing requires lean thinking to reduce waste and enhance business values. Integrating the previous sections in this chapter, the following eight steps can help you to develop "fat-free writing":

1. Circle the prepositions. Too many prepositions can drain all the action out of a sentence. Get rid of the prepositions, and find a strong active verb to make the sentence direct:

Original	Revised
In this report is an example of the use of the Minor's Rule in fatigue design.	This report exemplifies fatigue design using the Minor's Rule.

2. Circle the "is" forms. Using "is" in a sentence gets it off to a slow start and makes the sentence weak. Replace as many "to be" verbs with action verbs as you can, and change all passive voice ("is designed by") to an active voice ("designs").

Original	Revised
The point I wish to make is that failure develops with defects growing.	Failure develops with defects growing.

3. Ask, "Where's the action?" "Who's kicking who?" (Using Lanham's own terminology here—to be precise, it would be "Who kicks whom?"). If you get stuck in a passive sentence, always ask the question: "Who does what to whom?" If you use this formula, you will always write active sentences.

Original	Revised
Fatigue is considered accumulative damage by some people.	Some people consider fatigue accumulative damage.

4. Put this "kicking" action in a simple active verb.

Original	Revised
The theory of relativity is not used by this engineering design.	This engineering design does not use the theory of relativity.

5. Start fast—no slow windups. Stick to the action and avoid opening sentences with phrases such as these:

My opinion is that ...

The point I wish to make is that ...

The fact of the matter is that ...

6. In general, avoid starting sentences with "There is," "There are," or "There were":

Original	Revised
There are many ways in which we can classify engineering laboratories.	We can classify houses many ways.
There was a long line of new compressors on the final inspection machine.	New compressors lined the final inspection machine.

7. Remove "to be" whenever possible:

Original	Revised
Stephen Timoshenko was considered to be an excellent mechanical engineer.	Stephen Timoshenko was considered an excellent mechanical engineer.
Many engineers find the lectures to be stimulating.	Many engineers find the lectures stimulating.

8. Avoid verbal detours:

Original	Revised
It is essential that there be no new construction of houses in the area designated as the sanctuary for wildlife.	No new houses should be built in the wildlife sanctuary.

Bibliography

Alred, G. J., *Business Writer's Handbook*. 6th ed. New York: St. Martin's, 2000.

American Management Association, *The AMA Style Guide for Business Writing*. New York: AMACOM, 1996.

Beamer, L. and Varner, I. I., *Intercultural Communication in the Global Workplace*. 2nd ed. Boston: McGraw-Hill/Irwin, 2001.

Chaney, L. H. and Martin, J. S., *Intercultural Business Communication*. 2nd ed. Upper Saddle River, NJ: Prentice Hall, 2000.

Cialdini, R. B., *Influence: The Psychology of Persuasion*. Rev. ed. New York: Morrow, 1993.

Downey, R., Boland, S., and Walsh, P., *Communications Technology Guide for Business*. Boston: Artech House, 1998.

Eckhouse, B. E., *Competitive Communication: A Rhetoric for Modern Business*. Rev. ed. New York: Oxford University Press, 1999.

Fearn-Banks, K., *Crisis Communications: A Casebook Approach*. 2nd ed. Mahwah, NJ: Lawrence Erlbaum Associates, 2002.

Gardner, H., *Changing Minds: The Art and Science of Changing Our Own and Other Peoples Minds*. Boston: Harvard Business School Press, 2004.

Geffner, A. B., *How to Write Better Business Letters*. 3rd ed. Hauppauge, NY: Barron's, 2000.

Geffner, A. B. and NetLibrary, Inc., *Business English: A Complete Guide to Developing an Effective Business Writing Style*. 3rd ed. Hauppauge, NY: Barron's Educational Series, 1998.

Guilar, Joshua D. *The Interpersonal Communication Skills Workshop: Listening, Assertiveness, Conflict Resolution, Collaboration*. New York: AMACOM, 2001.

Harvard Business Essentials: Business Communication. Boston: Harvard Business School Press, 2003. (The Harvard Business Essentials series.)

Harvard Business Review on Effective Communication. Boston: Harvard Business School Press, 1999. (The Harvard Business Review paperback series.)

Holtz, S., *Corporate Conversations: A Guide to Crafting Effective and Appropriate Internal Communications*. New York: AMACOM, 2004.

Jablin, F. M. and Putnam, L., *The New Handbook of Organizational Communication: Advances in Theory, Research, and Methods*. Thousand Oaks, CA: Sage Publications, 2001.

Levine, R. V., *The Power of Persuasion: How We're Bought and Sold*. Hoboken, NJ: John Wiley & Sons, 2003.

McLeary, J. W., *By the Numbers: Using Facts and Figures to Get Your Projects and Plans Approved*. New York: American Management Association, 2000.

Munter, M., *Guide to Managerial Communication: Effective Business Writing and Speaking*. 4th ed. Upper Saddle River, NJ: Prentice Hall.

Ohno, T., *Toyota Production System: Beyond Large-Scale Production*. Portland, OR: Productivity Press, 1988.

Pan, Y., Scollon, S. B. K, and Scollon, R., *Professional Communication in International Settings*. Malden, MA: Blackwell Publishers, 2002.

Pearce, T., *Leading Out Loud: Inspiring Change through Authentic Communication*. New and rev. ed. San Francisco: Jossey-Bass Publishers, 2003.

Rankin, E., *The Work of Writing: Insights and Strategies for Academics and Professionals*. San Francisco: Jossey-Bass, 2001.

Ryan, K., *Write up the Corporate Ladder: Successful Writers Reveal the Techniques That Help You Write with Ease and Get Ahead*. New York: American Management Association, 2003.

Simmons, J., *We, Me, Them, & It: The Powers of Words in Business*. New York and London: Texere, 2002.

Stockard, O., *The Write Approach: Techniques for Effective Business Writing*. San Diego: Academic Press, 1999.

Tingley, J. C., *The Power of Indirect Influence*. New York: AMACOM, 2001.

Whalen, D. J., *I See What You Mean: Persuasive Business Communication*. Thousand Oaks, CA: Sage Publications, 1996.

Wiener, V., *Power Communications: Positioning Yourself for High Visibility*. New York: New York University Press, 1994.

Weissman, J., *Presenting to Win: Persuade Your Audience Every Time*. Upper Saddle River, NJ: Financial Times/Prentice Hall, 2003.

Worth, R., *Webster's New World Business Writing Handbook*. Indianapolis: Wiley Publishing, 2002.

9

Write Actively—Engineering Is about Actions

9.1 Active Voice: "Albert Einstein Wrote the Theory of Relativity"

Active writing motivates your engineering readers. Writing passively takes more words and often becomes wordy, roundabout, and confusing. You are using the passive voice in a sentence if the subject of the sentence is being acted upon instead of acting on its own. Passive voice obscures or loses part of the substance (i.e., the actor) of a sentence. The following sentence is written in the active voice:

> "Albert Einstein wrote the Theory of Relativity."

Now, consider the following revision of that sentence, written in passive voice:

> "The Theory of Relativity was written by Albert Einstein."

When you use passive voice, the receiver of the action becomes the subject of the sentence, and the actor appears in a prepositional phrase after the verb. The passive voice is used everywhere in the engineering literature, yet it is generally more difficult to use accurately, and to understand accurately, than the active alternative.

The verb is the most powerful transmitter in the sentence—it carries the energy and the image—and the sentence's message is pulled along by the verb. Active verbs render text more dynamic and more interesting to read; therefore, engineering writing should employ active verbs to capture readers' attention. Yet, engineers frequently use passive verbs in their writing to mask action and sidetrack readers' attention to objects instead of active agents. For example,

> "It was announced by the program manager that an SEM had been purchased for the surface defects of the thin films to be examined."

> Active writing minces no words. It calls attention directly to responsible agents, stating clearly what they do and how they affect their surroundings. Because of these, active writing motivates and engages readers for actions.

FIGURE 9.1
The power of active writing.

takes much longer to say than the following active version,

> "The program manager announced that we had purchased an SEM to examine the surface defects of the thin films."

Active writing also provides the power and precision for engineering actions. When you use the passive voice, you are choosing to give your reader only a certain amount of information (see Figure 9.1). And if you are making that choice because you do not have any more information to give, or because you do not know how to express that information, then you are not as in control of your sentence as you could be. This lack of control translates into sloppy, unclear prose. Your words must connect with your readers immediately, or else you will lose their attention—and their business.

9.2   How to Recognize the Passive Voice

Since the passive voice is so often a product of imprecision, many engineers work on weeding it out of their prose during the revision process. In sentences written in passive voice, the subject receives the action expressed in the verb; the subject is acted upon. The agent performing the action may appear in a "by the ... " phrase or may be omitted. If you can ask, "By whom?" or "By what?" after a verb, the verb is in the passive voice. When we write in the passive voice, we do four things:

1. We either delete the original subject or put it at the end of the sentence with "by" before it.
2. We transfer the direct object to the subject slot.

3. We substitute the past participle of the verb in the verb slot.

4. We add a form of the verb "be" as auxiliary to the past participle.

The more regularly you revise for a particular element of style, the more likely you will avoid misusing it in future first drafts. Active writing starts from identifying passive constructions in your writing. Try these steps for identifying passive constructions in your writing:

1. The technical signs of passive voices are any form of "to be" plus the past participle of a main verb. Scan your writing for any of the following:

   - A general overabundance of the use of the verb to be (is, are, was, were, be, am, being, been);
   - Verb phrases that include the verb "to be," such as "is written," "was discovered by," and "are being presented"
   - Sentences that begin with "It is," "There is," "This is," "What is," and "It was discovered that … ,"

2. Try switching the sentence into the active voice on a separate piece of paper. Sometimes this will be a simple process of locating the subject and placing it at the beginning of your sentence instead of at the end:

   A test sample was dropped on the table by the technician
   ⇒
   The technician dropped a test sample on the table.

   If your passive sentence does not have a subject, you will need to add one to make it active. This often indicates that your original sentence was unnecessarily vague:

   The accelerated life test was performed.
   ⇒
   The engineering laboratory performed the accelerated life test.

3. Ask yourself whether you need to present the sentence (or part of the sentence) in passive language. Would the meaning change significantly—for the worse—if you switched the structure around? Here is an example of a sentence that depends on a passive construction for its meaning:

   The memory chip was hit by a neutron last week and could not function in last night's reliability test.

   The most important agent in this sentence is the memory chip, not the neutron; changing the first clause around to "A neutron hit the memory chip last week" would not transition smoothly into the second clause, and would confusingly present the neutron as more meaningful than the memory chip that was hit. Most important, the changed version just would not sound right. Read your revisions out loud as a final means of checking for clarity.

If you have determined that your sentence does not need to be written in the passive voice, change it around. Any sentences that do need to be written passively will be all the more effective if they are surrounded by precise, active prose.

9.3   How to Write Actively—Use Three Cures

In engineering writing, overuse of passive voice or use of passive voice in long, complicated sentences can cause readers to lose interest or to become confused. Sentences in active voice are generally—though not always— clearer and more direct than those in passive voice. In sentences written in active voice, the subject performs the action expressed in the verb; the subject acts. Passive verbs cause problems. They make writing wordy, roundabout, and sometimes downright confusing. Learn how to spot passive verbs and make them active. Most sentences should use a "who-does-what order." By leading with the doer, you automatically avoid a passive verb. Use these three cures to avoid passive voices in engineering writing:

- Put a doer before the verb.
- Drop part of the verb.
- Change the verb.

9.3.1   Put a Doer before the Verb

Never use the passive where the active will do. Put the doer before the verb (e.g., "Sarah wrote the report" instead of "The report was written by Sarah." Try switching the sentence into the active voice on a separate piece of paper. Sometimes, this will be a simple process of locating the subject and placing it at the beginning of your sentence instead of at the end.

Passive	Active
The test equipment must have been misplaced by the truck carrier.	The truck carrier must have misplaced the test equipment.
The ship was inspected by a quality engineer.	A quality engineer inspected the ship.
Appropriate clothing will be worn by all personnel.	Wear appropriate clothing.

If your passive sentence does not have a subject, you will need to add one to make it active. This often indicates that your original sentence was unnecessarily vague.

Passive	Active
The test requested must be approved in advance.	The lab manager must approve the test request in advance.

9.3.2 Drop Part of the Verb

Sometimes, you can transform passive voice to active voice by dropping part of the verb. Remove the form of the "to be" verb from the sentence. Change the past participle into the appropriate tense. To determine the tense of the verb, look at the tense of the "to be" verb in the passive sentence.

Passive	Active
The motors are contained in the box.	The motors are in the box.
The results are listed in the attachment.	The results are in the attachment.
The project leader will be moved to a new office today.	The project leader will move to a new office today.
Then he was transferred to Advanced Engineering.	Then he transferred to Advanced Engineering.
The parts are listed on the label.	The list of parts is on the label.

9.3.3 Change the Verb

Changing the verb could also make your writing active. Verbs provide the momentum of writing. Proper verb choice often makes the difference between clear writing and clumsy writing.

Passive	Active
E-mail and memo writings are shown in Appendix A.	E-mail and memo writings appear in Appendix A.
The new SEM has not been received yet.	The new SEM has not arrived yet.
Laboratory personnel are prohibited from doing so.	Laboratory personnel must not do so.
The propeller shaft has not been received.	The propeller shaft has not arrived.

Use the active voice predominantly and the passive voice sparingly. It will help keep your writing crisp and your audience interested and pining for more.

9.4 Write Passively for Good Reasons Only

Should you try always to write in the active voice? No. You may have good reasons to highlight the object of the sentence, to emphasize the results, and to de-emphasize the actor. Here is a summary of the good reasons to use the passive voice.

Use passive voice ...	Example
To emphasize the action instead of the actor	After several meetings, the proposal was approved by the review board. Gold can be bent, melted, or hammered.
To keep the subject and focus consistent throughout a passage	The IT department recently presented what proved to be a controversial proposal to expand its data processing capability. After several meetings, the proposal was approved by ...
To be tactful by not naming the actor	The instructions were somehow misinterpreted.
To describe a condition in which the actor is unknown or unimportant	The laptop has been damaged. Every year, thousands of devices are found as having defects.
To create an authoritative tone	Tests are not allowed after 6:00 p.m.

The choice of active or passive voice is largely a matter of what you want to emphasize and what question you want to answer. Compare the following sentences:

Sentence	Emphasis	Question Answered
We used an SEM to examine the surface defects of the thin film coating	The actor—"We"	*Who used an SEM to examine the surface defects of the thin film coating?*
We examined the surface defects of the thin film coating using an SEM.	The actor—"We"	*Who examined the surface defects of the thin film coating using an SEM?*
An SEM was used to examine the surface defects of the thin film coating	The object—"An SEM"	*What instrument was used to examine the surface defects of the thin film coating?*
Surface defects of the thin-film coating were examined with an SEM.	The result—"Surface defects ... were examined ... "	*What were examined with an SEM?*

Write passively if you have good reason to avoid saying who or what has done the verb's action. You might do that when the doer is unknown, unimportant, obvious, or better left unsaid. When in doubt, write actively, even though the doer may seem obvious. Active writing is more direct and vigorous. It makes your writing briefer, more interesting, and easier to understand.

In summary, the nature of your engineering writing may require using the passive voice. Professional reports emphasize results and the objects of actions. You can certainly use the passive voice to vary the cadence or rhythm of a paragraph. In these cases, instead of worrying about voice, use strong verbs instead of weak verb phrases. Overall, though, you serve your readers better with the active voice, because it is clearer.

9.5 Theory of Completed Staff Work

Engineers are staff members of business organizations. Completed engineering work is based on the theory of "completed staff work." This standard has guided countless professionals since the 1940s, when an Army general sent the first version to his staff.

Engineering provides recommendations for business decisions. We refer to an engineering recommendation that has been properly prepared and coordinated as completed engineering work. Completed engineering work is a single proposed recommendation that:

- Has been thoroughly analyzed based on engineering principles
- Has been coordinated for business implementations
- Represents the best engineering recommendation possible
- Simply requires approval or disapproval
- Is prepared in final form for signature

The following guidelines will help you to accomplish a completed engineering work.

9.5.1 Conclude before Analyzing

Completed engineering work consists of studying a problem and presenting its solution in such a way that the boss need only indicate her or his approval or disapproval of the completed action. The concept, completed action, is emphasized because the more difficult a problem is, the more the tendency is to present the problem to the boss in a piecemeal fashion. It is the duty of an engineer to work out the details, no matter how perplexing they may be. The writing, whether it involves the pronouncement of a new product or affects an existing one, should be worked out in finished form before presentation to the boss for decision.

9.5.2 Present Answers, Not Questions

It is your job to advise your boss what she or he ought to do, not to ask your boss what you ought to do. The impulse that often comes to the inexperienced staff member—to ask the boss what to do—recurs more often when the problem is difficult. It is so easy to ask the boss what to do, and it appears so easy for her or him to answer. Resist that impulse. You will succumb to it only if you do not know your job. Your job is to tell the boss what she or he ought to do, not to ask her or him what you should do. The boss needs answers, not questions. Your job is to study, write, restudy, and rewrite until you have worked out a single proposed action—the best one of all the actions you have considered. The boss merely approves or disapproves. When presenting alternative courses of engineering actions, you should say which alternative you think is best.

9.5.3   Elaborate Only When Necessary

The boss should not be burdened with long explanations and memos. Writing a memo to the boss does not constitute completed engineering work, but writing a memo for the boss to send to someone else does. Your views should be placed before the boss in finished form, so that the boss can make them her or his views simply by signing the document. In most instances, completed engineering work results in a single document prepared for the signature of the boss, without accompanying comment. If the proper result is achieved, the boss will usually recognize it at once. If further comment or explanation is needed, the boss will ask for it.

9.5.4   Sell Your Ideas More Readily

The theory of completed staff work does not preclude a rough draft, but the rough draft must not be a half-baked idea. It must be complete in every respect, except that it lacks the requisite number of copies and need not be neat; however, a rough draft must not be an excuse for shifting to the boss the burden of formulating the action. The completed staff work concept may result in more work for the engineer, but it results in more freedom for the boss. This is as it should be. Further, it does two things:

1. The boss is protected from half-baked ideas, voluminous memos, and immature oral presentments.
2. The engineer who has a real idea to sell can find an application more readily.

9.5.5   Would You Sign the Document?

When you have finished your engineering work, the final test is this: If you were the boss, would you sign the document you have prepared and stake your professional reputation on its correctness? If the answer is no, take it back and rework it, because it is not yet completed engineering work.

9.5.6   Preparing Completed Staff Work

The theory of completed staff work simply recognizes that finding problems is easy. What is hard is recommending the appropriate solution(s) for a given project's resources, capabilities, schedule, and then implementing the solution(s) that result in improved project performance. Table 9.1 highlights

TABLE 9.1

Procedure for Preparing Completed Staff Work

Action	Description
Define the problem	The way things are
List criteria	The way things ought to be
Form assumptions	Conditions we cannot verify
Identify constraints	Conditions we cannot change
Develop alternatives	Ways that might solve the problem
Select best alternative	The clinching argument: • Outline of pros and cons • Consideration of nonconcurrences • A decision matrix if it will clarify the issue
Submit for approval	Best alternative, recommended to the decision maker and an explanation of why you chose it over the others

the procedure for preparing completed staff work, particularly an action that requires a signature or other means of approval.

The following checklist will help you implement the theory of completed staff work with a step-by-step process:

Ideas/Content (Development)

The writing is clearly focused, which leads to achieving a well-defined goal.

- The purpose is clearly defined.
- The writing supports the purpose with concise, logical details that meet the reader's informational needs.
- Sources, if used, are acknowledged.

Organization

The organization enhances and showcases the purpose. The sequence, structure, and presentation are compelling and move the reader through the text.

- Information is arranged in a format that is logical and effective and meets the reader's needs.
- The writing is a comprehensive piece with a constructive introduction, a body that provides relevant information, and a suitable conclusion that reinforces the purpose and leaves the reader with a sense of completion.
- Transitions are appropriate and connect the ideas.
- Information is organized within each section, paragraph, list, or graphic in a logical and effective sequence to meet the reader's needs.

Voice

The writer speaks directly to the reader in a way that is individualized, expressive, and engaging. Clearly, the writer is involved in the text and is writing for an audience.

- The text and graphics are appropriate for the audience and purpose (e.g., letter, complex reports, directions, brochures, electronic presentations, newsletters, memos, e-mails, fliers, Web pages, charts, maps, tables, pictorials, and resumes).
- Writes with authority, so the voice is not distracting.

Word Choice

Words convey the intended message in an accurate and concise manner that increases the reader's understanding.

- Words are clear, precise, and professional.
- The meaning of technical terms or professional jargon is defined or can be determined by the context.
- The vocabulary suits the purpose, subject, and audience.

Sentence Fluency

The text flows easily with a variety of sentence structures and lengths.

- Compact sentences or phrases make the point clear.
- The text reflects logic and sense, and helps show how ideas relate.
- Fragments, if used, work well.
- Dialogue, if used, is natural and convincing.

Conventions

The writer demonstrates control of standard writing conventions and uses them effectively to enhance readability. Errors tend to be so few and minor that the reader can easily skim right over them.

- Paragraph division is sound and reinforces the organizational structure.
- Grammar and usage are correct and contribute to clarity and style.
- Punctuation is smooth and guides the reader through the text.
- Spelling is generally correct, even on more difficult words.
- Only light editing would be required to polish the text for publication.
- Graphic devices, when used, are clear, helpful, visually appealing, and supportive of the text.
- The writer may manipulate conventions, particularly grammar, for stylistic effect.

Bibliography

Alred, G. J., *Business Writer's Handbook*. 6th ed. New York: St. Martin's, 2000.

American Management Association, *The AMA Style Guide for Business Writing*. New York: AMACOM, 1996.

Beamer, L. and Varner, I. I., *Intercultural Communication in the Global Workplace*. 2nd ed. Boston: McGraw-Hill/Irwin, 2001.

Chaney, L. H. and Martin, J. S., *Intercultural Business Communication*. 2nd ed. Upper Saddle River, NJ: Prentice Hall, 2000.

Cialdini, R. B., *Influence: The Psychology of Persuasion*. Rev. ed. New York· Morrow, 1993.

Downey, R., Boland, S., and Walsh, P., *Communications Technology Guide for Business*. Boston: Artech House, 1998.

Eckhouse, B. E., *Competitive Communication: A Rhetoric for Modern Business*. Rev. ed. New York: Oxford University Press, 1999.

Fearn-Banks, K., *Crisis Communications: A Casebook Approach*. 2nd ed. Mahwah, NJ: Lawrence Erlbaum Associates, 2002.

Gardner, H., *Changing Minds: The Art and Science of Changing Our Own and Other People's Minds*. Boston, MA: Harvard Business School Press, 2004.

Geffner, A. B., *How to Write Better Business Letters*. 3rd ed. Hauppauge, N.Y: Barron's, 2000.

Geffner, A. B. and NetLibrary, Inc., *Business English: A Complete Guide to Developing an Effective Business Writing Style*. 3rd ed. Hauppauge, NY: Barron's Educational Series, 1998.

Guilar, J. D., *The Interpersonal Communication Skills Workshop: Listening, Assertiveness, Conflict Resolution, Collaboration*. New York: AMACOM, 2001.

Harvard Business Essentials: Business Communication. Boston: Harvard Business School Press, 2003. (The Harvard Business Essentials series)

Harvard Business Review on Effective Communication. Boston: Harvard Business School Press, 1999. (The Harvard Business Review paperback series)

Holtz, S., *Corporate Conversations: A Guide to Crafting Effective and Appropriate Internal Communications*. New York: AMACOM, 2004.

Jablin, F. M. and Putnam, L., *The New Handbook of Organizational Communication: Advances in Theory, Research, and Methods*. Thousand Oaks, CA: Sage Publications, 2001.

Levine, R. V., *The Power of Persuasion: How We're Bought and Sold*. Hoboken, NJ: John Wiley & Sons, 2003.

McLeary, J. W., *By the Numbers: Using Facts and Figures to Get Your Projects and Plans Approved*. New York: American Management Association, 2000.

Munter, M., *Guide To Managerial Communication: Effective Business Writing and Speaking*. 4th ed. Upper Saddle River, NJ: Prentice Hall.

New, C. C., *How to Write a Grant Proposal*. Hoboken, NJ: John Wiley, 2002.

Ohno, T., *Toyota Production System: Beyond Large-Scale Production*. Portland, OR: Productivity Press, 1988.

Pan, Y., Scollon, S. B. K., and Scollon, R., *Professional Communication in International Settings*. Malden, MA: Blackwell Publishers, 2002.

Pearce, Terry. *Leading out Loud: Inspiring Change through Authentic Communication*. New and rev. ed. San Francisco: Jossey-Bass Publishers, 2003.

Rankin, E., *The Work of Writing: Insights and Strategies for Academics and Professionals.* San Francisco: Jossey-Bass Publishers, 2001.

Ryan, K., *Write up the Corporate Ladder: Successful Writers Reveal the Techniques That Help You Write with Ease and Get Ahead.* New York: American Management Association, 2003.

Simmons, J., *We, Me, Them, & It: The Powers of Words in Business.* New York and London: Texere, 2002.

Stockard, O., *The Write Approach: Techniques for Effective Business Writing.* San Diego: Academic Press, 1999.

Tingley, J C., *The Power of Indirect Influence.* New York: AMACOM, 2001.

Whalen, D. J., *I See What You Mean: Persuasive Business Communication.* Thousand Oaks, CA: Sage Publications, 1996.

Wiener, V., *Power Communications: Positioning Yourself for High Visibility.* New York: New York University Press, 1994.

Weissman, J., *Presenting to Win: Persuade Your Audience Every Time.* Upper Saddle River, NJ: Financial Times/Prentice Hall, 2003.

Worth, R., *Webster's New World Business Writing Handbook.* Indianapolis: Wiley Publishing, 2002.

3

Integrating Your Speaking and Writing Skills

10

Everyday Engineering Communications— E-Mails, Phone Calls, and Memos

---

E-mails, phone calls, and memos are important tools for everyday business communication. A phone call is real-time, immediate, and interactive, whereas a hard-copy memo appears concrete and official. E-mail is a cross between a phone call and a memo. Although not as interactive as a phone call, e-mail still allows you to reply to another e-mail easily and quickly. Although not as concrete as a memo, e-mail still provides a record that can be printed and stored.

10.1 Effective E-mail Writing: Seven Things to Remember

Electronic communication, because of its speed and broadcasting ability, is fundamentally different from paper-based communication. Because the turnaround time can be so fast, e-mail is more conversational than traditional paper-based media.

In a paper document, it is absolutely essential to make everything completely clear and unambiguous, because your audience may not have a chance to ask for clarification. With e-mail documents, your recipient can ask questions immediately. E-mail thus tends, in the same way as conversational speech, to be sloppier than communications on paper.

This is not always bad. It makes little sense to slave over a message for hours, making sure that your spelling is flawless, your words eloquent, and your grammar beyond reproach, if the point of the message is to tell your co-worker that you are ready to go to lunch.

However, your correspondent also will not have normal status cues, such as dress, diction, or dialect, and may make assumptions based on your name, address, and—above all—facility with language. You need to be aware of when you can be sloppy and when you have to be meticulous.

E-mail also does not convey emotions nearly as well as face-to-face or even telephone conversations. It lacks vocal inflection, gestures, and a shared environment. Your correspondent may have difficulty telling if you are serious or kidding, happy or sad, frustrated or euphoric. Sarcasm is particularly dangerous to use in e-mail.

Another difference between e-mail and older media is that what the sender sees when composing a message might not look like what the reader sees. Vocal cords make sound waves that are perceived basically the same by both the speaker's ears as the audience's. The paper on which a note is written is the same paper that the recipient sees. With e-mail, however, the software and hardware used for composing, sending, storing, downloading, and reading may be completely different from what the correspondent uses. A message's visual qualities may be quite different by the time it gets to someone else's screen.

Thus, your e-mail compositions should be different from both your paper compositions and your speech. This section describes how to tailor your message to this new medium.

10.1.1   Stop, Think, Then Write (or Don't!)

Are you e-mailing to say you will call when the fax goes through? Is a phone call more appropriate? Choosing the right communication medium will increase your chance of being heard.

10.1.2   Write an Informative and Engaging Subject Line

Recipients usually scan the subject line to decide whether to open, forward, file, or trash a message. Remember, your message is not the only one in your recipient's mailbox. Write an informative and engaging subject line so your message will be opened first.

For	Try
Some Thoughts.	How to Design an Efficient Power Supply
Meeting Schedule.	Meeting Rescheduled Due to Presentation to Sears.
Meeting—Important! Read Immediately!!	Do we need a larger room for design review meeting next Friday?

10.1.3   Keep the Message Focused and Readable

Position your key point up front on the screen. Often, recipients only read part of the way through a long message, hit "reply" as soon as they have something to contribute, and forget to keep reading. This is part of human nature.

If your e-mail contains multiple messages that are only loosely related, in order to reduce the risk that your reader will reply only to the first item that grabs his or her fancy, you could number your points to ensure they are all read (adding an introductory line that states how many parts are included in the message). If the points are substantial enough, split them up into separate messages so your recipient can delete, respond, file, or forward each item individually.

Keep your message readable.

- Use a combination of upper and lower case, white space, and a legible font (Times New Roman or Arial, for example).
- Use standard capitalization and spelling, *especially* when your message asks for support.
- Skip lines between paragraphs.
- Avoid fancy typefaces. Do not depend upon bold font or large size to add nuances—many people's e-mail readers only display plain text. In a pinch, use asterisks to show *emphasis.*
- Do not type in all caps. Online, all caps means shouting. Regardless of your intention, people will react as if you meant to be aggressive.

10.1.4   Identify Yourself Clearly

When contacting someone cold, always include your name, occupation, and any other important identification information in the first few sentences. Personalize by using contractions, pronouns, and a conversational tone.

If you are following up on a face-to-face contact, you might appear too timid if you assume your recipient does not remember you. However, you can drop casual hints to jog his or her memory: "I enjoyed talking with you about PDAs in the elevator the other day."

10.1.5   Keep Your Cool

Your emotional state can slip into an e-mail without notice, with curt sentences, skipped pleasantries, and blunt questions. If you find yourself writing in anger, take a break. Take some time to cool off before you hit "send."

10.1.6   Proofread

Take the time to make your message look professional. Remember, your recipient can easily forward your e-mail to someone else.

Although your spell checker will not catch every mistake, at the very least it will catch a few typos. If you are sending a message that will be read by someone higher up in the chain of command (e.g., a superior or professor), or if you are about to mass mail dozens or thousands of people, take an extra minute or two before you hit "send." Show a draft to a close associate to see whether it actually makes sense.

10.1.7   Wait a Moment before Hitting "Send"

Make sure your topic, tone, and style will not embarrass you. You lose control of your e-mail as soon as you hit "Send," so stick to professional language. To avoid sending a badly spelled, half written pile of rubbish, wait until you have written the e-mail before you key in the recipient's names. Hitting send too early is a painful, toe-curling experience.

10.2   How to Be Productive on the Phone

In today's business environment, a significant part of our working day is often spent on the telephone. To make this time most productive, specific telephone techniques can be applied daily.

10.2.1   Be Sharp and Professional

Remembering these points will help you to be sharp and professional in the way that you talk on the phone:

- *Have an aim:*

    When making an outgoing call, always know what you want to discuss. Always ensure that you have all the documentation you need to achieve your aim. This saves both your time and the time of the person to whom you are talking.

- *Tailor your style to that of the person to whom you are talking:*

    Busy people often prefer a clean-cut, direct approach with a bare minimum of social chat. Others may prefer a more sociable approach. Tailor your approach to the recipient's style.

- *Limit social conversation:*

    Social chat may be pleasant. If taken to extremes, however, it wastes time. It can be intensely frustrating if you have a lot of work to do.

- *Give concise answers to questions:*

    Long, rambling answers are unprofessional, dull, and confusing.

- *If you do not know an answer, say so:*

    If someone relies on you when you are guessing, and you guess wrong, then he or she will never trust you again. If you do not know something, say you will get back to that person with a firm answer.

- *At the end of a call, summarize the points made:*

    This ensures that both people agree on what has been said, and each person knows what action will be taken.

- *Do not talk to anyone else when on the phone:*

    This makes your organization look small. Put the other person on hold, then talk.

10.2.2   Three Distinct Stages

Each business phone call you make during the day can be thought of as having three distinct stages: a beginning, a middle, and an end. By properly handling each of these stages, you will maximize your chances for making a professional, effective presentation.

Several factors are critical at the beginning of a phone call. First, you should introduce yourself immediately so that the person with whom you are speaking knows who you are. Never simply say, "Hello" and expect that your voice will be recognized. Instead, say something like: "This is Robin Johnson, I'm calling in regard to … ." By giving your name and stating the reason for the call, you will establish a clearly defined beginning to the conversation. At the beginning, you should also establish that the person you are calling has the time to speak with you. If not, offer to call again at a more convenient time.

The middle stage of the conversation is the time to deal with the matter at hand. To best address this, be thoroughly prepared before you place the call. Organize your thoughts, perhaps jotting notes to lead you through the conversation and to avoid forgetting any important details. While speaking, be clear about each point, making sure that you have been understood. Do not sum up an important point by saying: "Do you know what I mean?" and assume that you have been understood. Always ask the other person to repeat the conclusion in his or her own words. This is the only way to make certain that you both come away from the conversation with mutual understanding.

The end of a conversation should be handled professionally to make sure that you have made a good impression during the call, and that all matters discussed have been covered and understood. Never lose track of time, allowing the call to stretch out meaninglessly. When all matters that need to be discussed have been concluded, you can review them. This is especially valuable if you or the person you called has specific tasks that need to be performed as a result of information discussed during the call. After all matters have been concluded, you can end the call simply by thanking the person for his or her time and hanging up after saying: "We'll speak again soon" or "Have a good day." If certain matters need further discussion in a future phone call, set a schedule for the call. In addition, be sure to determine who will call whom.

10.2.3   Managing Phone Time

Here are some additional time-saving tips that can help you manage your phone time:

1. When possible, block out a time to make all your phone calls, instead of scattering them throughout the day. This will help you build momentum and be more efficient.
2. Prioritize your calls, so that your most important calls are made first.

3. Set specific time limits for certain calls.

4. When leaving messages on voice mail, clearly and briefly state the reason for your call, when you will call back, or when you will be available should the person you called want to return your call.

5. When making several calls, be sure to take short breaks. Stand up, walk around, and take deep breaths. Stretch to avoid neck or shoulder tension. These activities will help keep you focused and relieve unnecessary fatigue and stress.

If you find yourself spending too much time during a workweek on the phone—and if phone work is taking you away from other important tasks—consult with a manager to see if your phone responsibilities can be eased.

10.3   "Memos Solve Problems"

Memos have one purpose in life. As the authors of *Business Writing Strategies and Samples* put it, "Memos solve problems."

Memos solve problems either by informing the reader about new information, such as policy changes and price increases, or by persuading the reader to take an action, such as attend a meeting, use less paper, or change a current production procedure. Regardless of the specific goal, memos are most effective when they connect the purpose of the writer with the interests and needs of the reader. This section will help you solve you memo-writing problems by discussing what a memo is, presenting some options for organizing memos, describing the parts of memos, and suggesting some hints that will make your memos more effective.

A memo is a no-nonsense professional document that is designed to be read quickly and passed along rapidly, often within a company or work group. E-mail messages are by far the most common form of memo. A business memo helps members of an organization communicate without the need for time-consuming meetings. It is an efficient way to convey routine information within an organization.

The writing style of a business memo is more formal that an e-mail, but it does not have to sound intimidating. Here, you want to effectively communicate your purpose to your engineering readers. When planning your memo, be sure to think about it from your reader's perspective. Pretend you are the recipient and ask yourself:

1. How is this relevant to me?
2. What, specifically, do you want me to do?
3. What's in it for me?

10.3.1 Keep Your Memo Structured

As with all writing, memo writing needs a structure. Because a memo is short, rambling meanderings will soon destroy its effectiveness and become a waste of productive time not only to those who read it, but also to the person who wrote it. As depicted below, a memo has a specific format.

Memo

Date:

To:

From:

Subject: Use the subject line to convey crucial details.

Opening Segment

Tell the reader why the memo has been written. Come to the point first—a business memo almost always begins with a bottom-line statement. A bottom-line statement or message is a short, terse statement of the memo's purpose.

Discussion Segment

Present details about the problem. Give enough information for decision makers to resolve the problem. Describe the task or assignment with details that support your opening paragraph (problem).

End with a Call to Action

Unless the purpose of the memo is simply to inform, you should finish with a clear call for action. Who should do what, and how long do they have to do it? You may need to include alternatives, in the event that your readers disagree with you. Be polite when you ask others to do work for you, especially when they are not under your supervision. You may wish to mention the actions that you plan to take next and what your own deadlines are, so your reader can gauge how important the project is to you.

Memo

Date: 10/28/2007

To: Jim Schneider, Manufacturing Engineer

From: Brian Lafferty, Project Engineer

Subject: Preliminary Design of Vibration Measurement and
          Display System

We need to implement the new product development plan sooner than discussed at our last meeting, because ABC company now has a similar one set to launch in November. Please review the product's manufacturability for lean manufacturing.

The preliminary design presented in the attached report uses the high-speed analog-to-digital (HSAD) converter and the network subsystems. The hardware added for the preliminary design includes a vibration measurement circuit connected to one of the high-speed A/D input pins on Port C and light emitting diodes (LEDs) connected to each pin of Port E. These LEDs act as vibration indicators. Additionally, the design includes a connection between the HSAD and a remote personal computer (PC) using an Ethernet connection. An assembly software program developed for this design performs built-in tests for using the added hardware. Jim, call me as soon as you have read the attached report.

10.3.2   Keep Your Memo Short and to the Point

Make your point quickly and succinctly, and give only the necessary information. If you have something longer than a page, it is better to send it as an attachment or a document that follows the memo used as a cover letter. Never make a memo too long. If someone takes a glance at a memo that appears to be too long, it will most likely be set aside for a time when that person is not busy. This can undermine the timeliness of your communication.

When writing a memo, use the following test to make sure it is to the point. Does the information stand alone without the use of headings and subheadings? Read the content without referring to the subject line. If it is understandable without this heading, then it is likely to be a good memo. If it is not clear, then something has probably been left out of the first sentence. When writing brief messages, it is tempting to take too much knowledge on the part of the reader for granted. You must think carefully about how your message will be received when writing a memo.

Bibliography

Alred, G. J., *Business Writer's Handbook*. 6th ed. New York: St. Martin's, 2000.

American Management Association, *The AMA Style Guide for Business Writing*. New York: AMACOM, 1996.

Beamer, L. and Varner, I. I., *Intercultural Communication in the Global Workplace*. 2nd ed. Boston: McGraw-Hill/Irwin, 2001.

Chan, J. F., *E-Mail: A Write It Well Guide—How to Write and Manage E-Mail in the Workplace*. Oakland, CA: Write It Well, 2005.

Chaney, L. H. and Martin, J. S., *Intercultural Business Communication*. 2nd ed. Upper Saddle River, NJ: Prentice Hall, 2000.

Cialdini, R. B., *Influence: The Psychology of Persuasion*. Rev. ed. New York: Morrow, 1993.

Downey, R., Boland, S., and Walsh, P., *Communications Technology Guide for Business*. Boston: Artech House, 1998.

Eckhouse, B. E., *Competitive Communication: A Rhetoric for Modern Business*. Rev. ed. New York: Oxford University Press, 1999.

Fearn-Banks, K., *Crisis Communications: A Casebook Approach*. 2nd ed. Mahwah, NJ: Lawrence Erlbaum Associates, 2002.

Gardner, H., *Changing Minds: The Art and Science of Changing Our Own and Other People's Minds*. Boston, MA: Harvard Business School Press, 2004.

Geffner, A. B., *How to Write Better Business Letters*. 3rd ed. Hauppauge, N.Y: Barron's, 2000.

Geffner, A. B. and NetLibrary, Inc., *Business English: A Complete Guide to Developing an Effective Business Writing Style*. 3rd ed. Hauppauge, NY: Barron's Educational Series, 1998.

Guilar, J. D., *The Interpersonal Communication Skills Workshop: Listening, Assertiveness, Conflict Resolution, Collaboration*. New York: AMACOM, 2001.

Harvard Business Essentials: Business Communication. Boston: Harvard Business School Press, 2003. (The Harvard Business Essentials series)

Harvard Business Review on Effective Communication. Boston: Harvard Business School Press, 1999. (The Harvard Business Review paperback series)

Holtz, S., *Corporate Conversations: A Guide to Crafting Effective and Appropriate Internal Communications*. New York: AMACOM, 2004.

Jablin, F. M. and Putnam, L., *The New Handbook of Organizational Communication: Advances in Theory, Research, and Methods*. Thousand Oaks, CA: Sage Publications, 2001.

Levine, R. V., *The Power of Persuasion: How We're Bought and Sold*. Hoboken, NJ: John Wiley & Sons, 2003.

McLeary, J. W., *By the Numbers: Using Facts and Figures to Get Your Projects and Plans Approved*. New York: American Management Association, 2000.

Munter, M., *Guide To Managerial Communication: Effective Business Writing and Speaking*. 4th ed. Upper Saddle River, NJ: Prentice Hall.

Ohno, T., *Toyota Production System: Beyond Large-Scale Production*. Portland, OR: Productivity Press, 1988.

Pan, Y., Scollon, S. B. K., and Scollon, R., *Professional Communication in International Settings*. Malden, MA: Blackwell Publishers, 2002.

Pearce, Terry. *Leading out Loud: Inspiring Change through Authentic Communication*. New and rev. ed. San Francisco: Jossey-Bass Publishers, 2003.

Rankin, E., *The Work of Writing: Insights and Strategies for Academics and Professionals*. San Francisco: Jossey-Bass Publishers, 2001.

Ryan, K., *Write up the Corporate Ladder: Successful Writers Reveal the Techniques That Help You Write with Ease and Get Ahead*. New York: American Management Association, 2003.

Simmons, J., *We, Me, Them, & It: The Powers of Words in Business*. New York and London: Texere, 2002.

Stockard, O., *The Write Approach: Techniques for Effective Business Writing*. San Diego: Academic Press, 1999.

Tingley, J C., *The Power of Indirect Influence*. New York: AMACOM, 2001.

Whalen, D. J., *I See What You Mean: Persuasive Business Communication*. Thousand Oaks, CA: Sage Publications, 1996.

Wiener, V., *Power Communications: Positioning Yourself for High Visibility*. New York: New York University Press, 1994.

Weissman, J., *Presenting to Win: Persuade Your Audience Every Time*. Upper Saddle River, NJ: Financial Times/Prentice Hall, 2003.

Worth, R., *Webster's New World Business Writing Handbook*. Indianapolis: Wiley Publishing, 2002.

11

Visuals for Engineering Presentation—
Engineers Think in Pictures

Presentations are an integral part of engineering. Engineers make presentations to propose projects, update progresses, and state results. Every time you make a presentation, you place your professional reputation on the line. In other words, you go "on stage" before your colleagues and critics. Much writing goes into a presentation, particularly in the creation of visuals, which can be computer projections or overhead transparencies.

How do you create effective visuals including slides, graphics, tables, and diagrams? If you want to design a technical visual that provides new learning for an engineering reader, what do you need to consider? Designing effective technical visuals requires you to understand some basic concepts about why technical visuals work:

- Vision is our dominant sense—seeing is believing.
- Readers remember visuals more quickly.
- Visuals help us in critical thinking and problem solving.
- Visuals promote efficient reading.
- Visuals do better than text at explaining spatial concepts.
- Visuals can reach international readers.

Human vision can quickly absorb large amounts of very precise data. The act of seeing, however, is not simply data gathering; seeing is a form of thinking. A reader can perceive a great deal of information visually, but she or he can only absorb a limited amount at once. A reader will search for a focal point in the visual. Therefore, you need to think carefully about how to create that focal point, and edit extraneous details from the visual without compromising the depth of content that the reader requires.

11.1 Optimize Slide Layout

A reader expects conventional patterns to help her or him in the effort to focus within and to analyze the visual. Basic layout principles help the reader: arrangement, proximity, alignment, and contrast. A reader also depends on the writer's/designer's ability to integrate labels, to use color skillfully, and to provide levels of pertinent information.

One advantage of using slides is that audiences remember more when the slides are well designed. For a technical presentation, you should set high goals for the presentation slides:

- Slides should help the audience during the talk.
- Slides should serve as notes for the audience after the talk.
- Slides should assist colleagues who have to give the same talk.

Slides might overpower a talk and draw attention to the actual visuals, causing your talk to fail. Effective slides *complement* a talk. They provide key text points that you will elaborate (never whole chunks of text that you read), or they contain a simplified graphic to give a visual image for a key concept. Two common errors made in the design of slides are:

- Creating slides that no one reads
- Creating slides that no one remembers

Four criteria are important in evaluating a layout design for presentation slides:

1. How readable is the design?
2. How memorable is the design?
3. How many slides does the design require?
4. Does the design help the slides stand as notes?

11.1.1   Create Slides That the Audience Can Read

One common error is having a slide format that dissuades the audience from reading. To avoid this error, an easily read typeface and layout are needed. Effective slides have the following:

A Strong Headline

The sentence headline should state succinctly the purpose or assertion of the slide. A strong headline:
- Identifies the slide's purpose for the audience
- Identifies the slide's purpose for you as the speaker

A Concise Slide Body

The body of a slide should support the headline primarily with images and with words where needed. A concise slide body:
- Primarily supports headline with images
- Supports with necessary words

Computer-Aided Design of Bridge

- A bridge can be reduced
  to a simple triangle.
- Strength of the bridge
  should be much greater
  than total load including
  - Weight of the bridge;
  - Weight of the cars;
  - "Wind load."

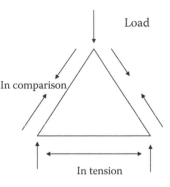

FIGURE 11.1
A sample slide.

Horizontal Format and Consistent Border

Actually, most overheads and screens are virtually square, so do not try to cram to the edges of a horizontal slide (see Figure 11.1).

Good Brightness and Contrast

Black on white always works well. A limited number of colors, perhaps three, can also be effective.

Letters That Are at Least a 20-Point Font

This goes for diagrams as well as text.

Clear, Simple Font

For slides, Arial or Helvetica are better than Times New Roman because they have letters that are all of equal width.

1 to 2 Minutes per Visual

Many engineering professionals try to "deal" overheads like they are dealing cards at a casino. Do not do this. Your audience needs time to absorb each visual.

11.1.2   Create Slides That the Audience Can Remember

A second common error is showing slides that the audience can read but does not remember. A reader who views the visual for reflection and learning tends to see the technical visual analytically. She or he is instinctively ranking information according to its importance to her or him. To make slides memorable, you have to consider what to include and what to exclude. Slides should include:

- Key results
- Key images

A reader approaches a technical visual within context and is influenced by her or his previous experience with technical visuals, level of knowledge, and by understanding the text that precedes and follows the visual. Because of these elements, slides should also include signals for the organization of the presentation:

- Beginning—title and purpose of the presentation
- Middle—discussion, analysis, and explanation
- Ending—conclusion, recommendation, and action items

In summary, a strong slide layout design presents the following benefits for presentation:

- The design is memorable for the audience.
- The design requires fewer slides, resulting in better pacing of the presentation.
- The design produces notes that stand alone.
- The design creates a more compelling argument.

11.2   Display Engineering Data Effectively

Graphics provide illustrated information to readers. In general, graphics are designed to make it easier for readers to understand your writing. Deciding when to insert a graphic depends on the information you need to convey. For example, as you are writing, you find yourself struggling to describe a complex concept. Fitting your description within a few paragraphs is impossible, so you decide to create a graphic. Often, graphics are useful when concepts, designs, or processes are too complex or cumbersome to describe in text. Why are graphics important in engineering writing? Because:

- Graphics are visually appealing.
- Graphics are easy to understand and remember.
- Graphics are indispensable for illustrating some types of relationships.
- Graphics serve as a universal language for your international readers.

Most professional, technical writing contains graphics—drawings, diagrams, photographs, illustrations of all sorts, tables, pie charts, bar charts, line graphs, flow charts, and so on. You can use graphics to represent the following elements in your technical writing.

11.2.1   Show Objects with Photographs, Drawings, and Diagrams

Engineers often think in pictures. If you are describing a fuel-injection system, you will probably need a drawing or diagram of the system. If you are explaining how to assemble a refrigerator door, you will need some illustrations of how that task is done. Photographs, drawings, and diagrams are the types of graphics that show objects. Here is a summary for effective photographs, drawings, and diagrams.

Photographs	Choose a photograph so the reader can see how an object looks
	Focus the photograph so that the important elements are visible, and extraneous detail is cropped away.
	Label the photograph so that readers understand what they are viewing.
	Check the photograph to be sure that representations of people or objects are not gender-biased or offensive.
Drawings	When a reader needs to see the parts of an object, choose a drawing.
	A drawing also can show a reader how to do a task more effectively by indicating position and relationship.
	Several types of drawings can be included: external, cutaway, cross section, exploded.
	Create the drawing from the perspective your reader will share.
	Focus on the key information you want to convey.
	Label the drawing so that readers understand its parts.
	Try to integrate labels into the drawing when possible.
Process Diagrams, CAD Diagrams, and Flowcharts	Diagrams can reveal abstract processes.
	Flowcharts show complex relationships that may be invisible.
	Use conventional technical symbols when appropriate.
	Decide what element of the process is most important.
	Explain and interpret the process diagram for your reader.
	Arrange activity so that the reader's attention works from left to right or from top to bottom.
	In flowcharts, place labels inside boxes to represent those activities.

Often technical information is most concisely and clearly conveyed by means other than words. Imagine how you would describe a simple triangle bridge accurately with words instead of the CAD diagram depicted in Figure 11.2.

If you direct the reader to a diagram, table, or equation, explain in plain words in the text what you want the reader to take from this. Do not assume that the reader will draw the conclusion you want him or her to draw. For diagrams, remember the following guidelines:

- Keep them simple.
- Draw them specifically for the product.
- Put small diagrams after the text reference and as close as possible to it.
- Think about where to place large diagrams.
- Be careful to describe the main message from diagrams.

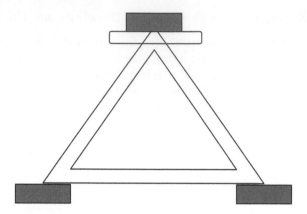

FIGURE 11.2
Simple CAD diagram for a triangle bridge.

Describe the main message from a diagram within the text so that the reader can continue without having to actually look at the diagram. Do not expect the reader to be able to break away from the text, interpret the data in the diagram, and then come back to the point you were making in the text without having forgotten what you were saying. A diagram should be viewed as supporting evidence or a helpful recasting of your arguments, to which the reader can refer if he or she needs convincing.

11.2.2   Show Numeric Data with Tables, Bar Charts, and Line Graphs

If you are discussing the rising cost of housing in Austin, Texas, you could use a table with the columns depicting five-year periods since 1970; the rows could be for different types of housing. You could show the same data in the form of bar charts, pie charts, or line graphs. Tables, bar charts, pie charts, and line graphs are some of the principal ways to show numerical data. Here is a summary for effective tables, bar charts, and line graphs.

Tables   Tables are effective when a reader needs to find particular pieces of data. A table allows him or her to find the desired fact(s).
Arrange the rows and columns of your table to support your reader's task.
Emphasize key information.
Label the columns and rows clearly. Indicate units measured.
Keep the table on one page, if possible.
Give acknowledgements of sources in table. Label and title table.

Bar Charts   Bar charts are effective when a reader needs to see relative proportions and amounts.
Draw bar charts to scale to help readers understand quantities.
Put a precise figure on top of bar when it is important for a reader to know both the precise number as well as to compare trends.
Use a background grid to help readers read across the bars.
Avoid three-dimensional bars because they easily distort information.
Be sure to label bar chart accurately.

Line Graphs	Line graphs depict trends and display relationships between two or more sets of data.
	Indicate the axes for a line graph clearly. The x-axis usually represents time. Start the axes at zero. If you cannot, then indicate with hash marks that the scale is not continuous from zero.
	Put labels next to lines and data points.
Pie Charts	Show simple comparisons, especially changes in quantity.
	Limit the number of bars.
	Be sure comparisons are clear.
	Adjust bar widths and space between them to make them equal.
	Arrange the order of bars carefully.
	Make creative choices.

Charts are actually just another way of presenting the same data that is presented in tables—although a more dramatic and interesting one. At the same time, however, you get less detail or less precision in a chart than you do in a table. Imagine the difference between a table of sales figures for a 10-year period and a line graph for that same data. You get a better sense of the overall trend in the graph but not the precise dollar amount.

Charts help a reader compare trends or volume. The most commonly used charts are bar charts, line graphs, and pie charts. Visuals composed of variations among small multiples are effective as well. Small multiples are sets of small graphics on a single page that represent aspects of a simple engineering feature.

11.2.2.1   Bar Charts

This type of graphic depicts simple comparisons, especially changes in quantity:

1. Limit the number of bars.
2. Be sure comparisons are clear.
3. Adjust bar widths and space between them to make them equal.
4. Arrange the order of bars carefully.
5. Make creative choices.
6. Use a background grid to help readers read across the bars.
7. Avoid three-dimensional bars because they easily distort information.
8. Be sure to label bar chart accurately.

Remember the following when developing a bar chart (see Figure 11.3).

- Draw bar charts to scale to help readers understand quantities.
- Put a precise figure on top of bar when it is important for a reader to know both the precise number as well as to compare trends.

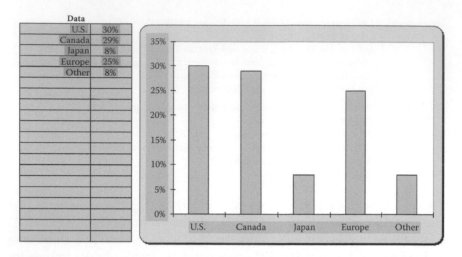

FIGURE 11.3
Bar chart for refrigerator sales.

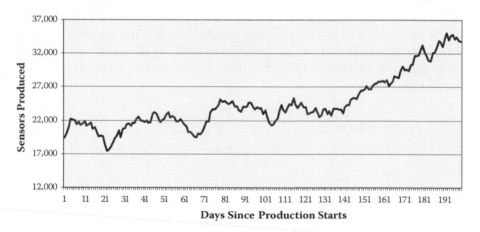

FIGURE 11.4
Line chart for sensor daily productions.

11.2.2.2 Line Graphs

This type of graphic is used to show trends or changes over time, such as price changes:

1. Show trends with line graphs.
2. Place line graphs where they can get attention.
3. Make line graphs that are accurate and clear.
4. Avoid putting numbers on the line graphs itself.
5. Do not place too many lines on the graphs.

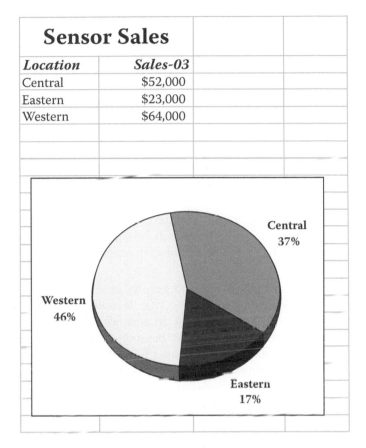

Sensor Sales			
Location	*Sales-03*		
Central	$52,000		
Eastern	$23,000		
Western	$64,000		

FIGURE 11.5
Pie chart for sensor daily productions.

6. Indicate the axes for a line chart clearly (see Figure 11.4). The x-axis usually represents time. Start the axes at zero. If you cannot, then indicate with hash marks that the scale is not continuous from zero.

7. Put labels next to lines and data points.

11.2.2.3 Pie Charts

Pie charts create a visual representation that helps engineers communicate the relationships and meanings of numbers. Pie charts are especially useful in representing proportions, percents, and fractions. When developing pie charts, remember the following:

- Label wedges of a pie chart clearly and indicate its percentage of the whole.
- Limit the number of wedges. Very small amounts could be illustrated more effectively in a different sort of visual.

- When using small multiples to indicate a trend over time, be sure to label them clearly and consistently.

11.2.2.4 Formatting Requirements

When you create charts, keep these requirements in mind (most of these elements are illustrated next and in Figure 11.5):

- Axis labels—In bar charts and line graphs, do not forget to indicate what the x- and y-axes represent. One axis might indicate locations, countries, or millions of dollars; the other, five-year segments from 1960 to the present.
- Keys—Bar charts, line graphs, and pie charts often use special colors, shading, or line style (solid or dashed). Be sure to indicate what these mean; translate them in a key (a box) in some unused place in the chart or graph.
- Figure titles—For most charts and graphs, you will want to include a title—in many cases, a numbered title. Readers need some way of knowing what they are looking at. In addition, do not forget to cite the source of any information you borrowed to create the graphic. The standard rule for when to number figures or tables is this: if you cross-reference the figure or table elsewhere in the text.

11.3   How to Develop Effective Graphics

As an engineer, you are allowed to avoid words entirely in certain places; diagrams are often much better than written text. Whole reports can be written with diagrams almost exclusively, and you should always consider using a diagram in place of a paragraph. Not only do diagrams convey some information more effectively, but they often assist in the analysis and interpretation of the data. For instance, a pie chart gives a quicker comparison than a list of numbers; a simple bar chart is far more intelligible than the numbers it represents. The only problem with diagrams is the writer often places less effort in their design than their information-content merits—and so some information is lost or obscured. Diagrams must be given due care: Add informative labels and titles, highlight any key entries, and remove unnecessary information. To develop effective graphics, remember the following:

- An effective graphic is appropriate to the writing situation.
- An effective graphic is self-explanatory.
- An effective graphic is placed correctly.
- An effective graphic is integrated with the text.

Bibliography

Alred, G. J., *Business Writer's Handbook*. 6th ed. New York: St. Martin's, 2000.

American Management Association, *The AMA Style Guide for Business Writing*. New York: AMACOM, 1996.

Arnheim, Rudolf. *Visual Thinking*. London: Faber and Faber, 1969.

Barry, A. M. S., *Visual Intelligence: Perception, Image and Manipulation in Visual Communication*. Albany, NY: State University Press, 1997.

Beamer, L. and Varner, I. I., *Intercultural Communication in the Global Workplace*. 2nd ed. Boston: McGraw-Hill/Irwin, 2001.

Chan, J. F., *E-Mail: A Write It Well Guide—How to Write and Manage E-Mail in the Workplace*, Oakland, CA: Write It Well. 2005.

Chaney, L. H. and Martin, J. S., *Intercultural Business Communication*. 2nd ed. Upper Saddle River, NJ: Prentice Hall, 2000.

Cialdini, R. B., *Influence: The Psychology of Persuasion*. Rev. ed. New York: Morrow, 1993.

Downey, R., Boland, S., and Walsh, P., *Communications Technology Guide for Business*. Boston: Artech House, 1998.

Eckhouse, B. E., *Competitive Communication: A Rhetoric for Modern Business*. Rev. ed. New York: Oxford University Press, 1999.

Fearn-Banks, K., *Crisis Communications: A Casebook Approach*. 2nd ed. Mahwah, NJ: Lawrence Erlbaum Associates, 2002.

Gardner, H., *Changing Minds: The Art and Science of Changing Our Own and Other People's Minds*. Boston: Harvard Business School Press, 2004.

Geffner, A. B., *How to Write Better Business Letters*. 3rd ed. Hauppauge, N.Y.: Barron's, 2000.

Geffner, A. B. and NetLibrary, Inc., *Business English: A Complete Guide to Developing an Effective Business Writing Style*. 3rd ed. Hauppauge, NY: Barron's Educational Series, 1998.

Guilar, J. D., *The Interpersonal Communication Skills Workshop: Listening, Assertiveness, Conflict Resolution, Collaboration*. New York: AMACOM, 2001.

Harvard Business Essentials: Business Communication. Boston: Harvard Business School Press, 2003. (The Harvard Business Essentials series)

Harvard Business Review on Effective Communication. Boston: Harvard Business School Press, 1999. (The Harvard Business Review paperback series)

Hilligoss, S., *Visual Communication: A Writer's Guide*. New York: Addison Wesley Longman, 2000.

Holtz, S., *Corporate Conversations: A Guide to Crafting Effective and Appropriate Internal Communications*. New York: AMACOM, 2004.

Jablin, F. M. and Putnam, L., *The New Handbook of Organizational Communication: Advances in Theory, Research, and Methods*. Thousand Oaks, CA: Sage Publications, 2001.

Kostelnick, C. and Roberts, D. D., *Designing Visual Language: Strategies for Professional Communicators*. Needham Heights, MA: Allyn & Bacon, 1998.

Levine, R. V., *The Power of Persuasion: How We're Bought and Sold*. Hoboken, NJ: John Wiley & Sons, 2003.

McLeary, J. W., *By the Numbers: Using Facts and Figures to Get Your Projects and Plans Approved*. New York: American Management Association, 2000.

Munter, M., *Guide To Managerial Communication: Effective Business Writing and Speaking*. 4th ed. Upper Saddle River, NJ: Prentice Hall.

Ohno, T., *Toyota Production System: Beyond Large-Scale Production*. Portland, OR: Productivity Press, 1988.

Pan, Y., Scollon, S. B. K., and Scollon, R., *Professional Communication in International Settings*. Malden, MA: Blackwell Publishers, 2002.

Pearce, Terry. *Leading out Loud: Inspiring Change through Authentic Communication*. New and rev. ed. San Francisco: Jossey-Bass Publishers, 2003.

Rankin, E., *The Work of Writing: Insights and Strategies for Academics and Professionals*. San Francisco: Jossey-Bass Publishers, 2001.

Ryan, K., *Write up the Corporate Ladder: Successful Writers Reveal the Techniques That Help You Write with Ease and Get Ahead*. New York: American Management Association, 2003.

Simmons, J., *We, Me, Them, & It: The Powers of Words in Business*. New York and London: Texere, 2002.

Stockard, O., *The Write Approach: Techniques for Effective Business Writing*. San Diego: Academic Press, 1999.

Tingley, J C., *The Power of Indirect Influence*. New York: AMACOM, 2001.

Tufte, E., *Envisioning Information*. Cheshire, CT: Graphics Press, 1990.

Tufte, E., *The Visual Display of Quantitative Information*. Cheshire, CT: Graphics Press, 1983.

Tufte, E., *Visual Explanation*. Cheshire, CT: Graphics Press, 1997.

Whalen, D. J., *I See What You Mean: Persuasive Business Communication*. Thousand Oaks, CA: Sage Publications, 1996.

Wiener, V., *Power Communications: Positioning Yourself for High Visibility*. New York: New York University Press, 1994.

Weissman, J., *Presenting to Win: Persuade Your Audience Every Time*. Upper Saddle River, NJ: Financial Times/Prentice Hall, 2003.

Worth, R., *Webster's New World Business Writing Handbook*. Indianapolis: Wiley Publishing, 2002.

12

Write Winning Grant Proposals

A grant proposal is not only a plan for solving a problem, but is also a business plan. To be successful, your grant proposal must sell both you and your technical ideas. The grant proposal is one of the most important forms of engineering writing. Successful proposals lead to funding for product innovation and process improvement.

12.1 Know Your Audience

A grant proposal is nothing more than an exercise in persuasive writing, and the key to being persuasive is to understand your audience. The audience for a grant proposal usually includes both business managers and engineers, and they view proposals in different ways. For instance, business managers review proposals to see if the plan for solving the problem is cost effective. They tend to be cautious people who like to take minimal risks and get good returns on their investments. Engineers, on the other hand, review proposals to see if the plan is technically feasible and innovative.

It is correct to assume that your readers are busy, impatient, and skeptical people, who have no reason to give your proposal special consideration and are faced with many more requests than they can grant, or even read thoroughly. Such a reader wants to find out quickly and easily the answers to these questions:

- What do you want to do, how much will it cost, and how much time will it take?
- How does the proposed project relate to the sponsor's interests?
- What will we learn as a result of the proposed project that we do not already know?
- What difference will the project make to: your organization, your employees, your discipline, the state, the nation, the world, or whatever the appropriate categories are?
- Why is it worth knowing and funding?
- What has already been done in the area of your project?
- What research and development have been done in a similar field?

- How will we know that the conclusions are valid?
- Why should you, instead of someone else, do this project?

Writing a grant proposal that answers these questions effectively requires considerable effort. Keep in mind that the reviewer may not read every word of your proposal. She or he may only read the abstract, the sections on research design and methodology, the vitae, and the budget. Make these sections as clear and straightforward as possible.

12.2   Understand Your Goal and Marketing Strategy

To sell your technical ideas successfully, your grant proposal needs to demonstrate three things:

1. You can do the research.
2. You have useful, creative ideas.
3. You are highly motivated to complete the work proposed.

Here, you must still persuade the audience that your technical idea is sound. Convince the audience that you have thought through the problem and have a workable solution. You need to explain the problem clearly and provide full background to give context to your solution.

In defining your goal, you need to think about the client's needs and requirements and your objectives. You need to convince the client that you understand his or her needs and can meet his or her requirements. Unlike the kind of proposal you would complete in a routine job, you also need to convince the client that you have the credibility, experience, and qualifications to do the job. As presented in Figure 12.1, you need to design your own marketing strategy.

"Marketing" makes many engineers uncomfortable because they think good science alone should convince potential buyers; however, you must still persuade the client that your technology is sound. Convince the client that you have thought through the problem and have a workable solution.

12.3   Select the Correct Writing Style

The way you write your grant proposal will tell the reviewers quite a bit about you. From reading your proposal, the reviewers will form an idea of who you are as a scholar, researcher, and as a person. They will decide whether

> Proposal development is a process for technical marketing, which brings together buyers and sellers for engineering products and services. A successful proposal convinces the potential buyers that
>
> - The proposal is sound technically and economically;
> - The writers are well qualified to accomplish the project.

FIGURE 12.1
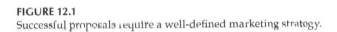
Successful proposals require a well-defined marketing strategy.

you are creative, logical, analytical, up-to-date in the relevant literature of the field, and, most important, capable of executing the proposed project. Allow your discipline and its conventions to determine the general style of your writing, and allow your own voice (and personality) to come through. Be sure to clarify your project's theoretical orientation and experimental foundation with data displays and diagrams.

Because most proposal writers seek funding from several different agencies or granting programs, it is a good idea to begin by developing a general grant proposal and budget. This general proposal is sometimes called a "white paper." It should explain your project to the engineering community. Before you submit proposals to different grant programs, you must tailor a specific proposal to each program's guidelines and priorities.

12.4 Organize Your Proposal around the Four Ps

To market your proposal, your need to organize it around the following four Ps:

- Product: The product aspects of marketing deal with the specifications of the actual goods or services that you are proposing. The scope of a product usually includes supporting elements such as merits, warranties, guarantees, and technical support.

- Pricing: This refers to the process of setting a price/cost for the product. The price/cost need not be monetary; it can simply be what is exchanged for the product or services (e.g., time, energy, psychology, or attention).
- Promotion: This includes publications/publicity, patents, and personal selling. It refers to the various methods of promoting the product and the project team.
- Placement (or distribution): This refers to when and how the product gets to the clients (e.g., the project schedule, principal investigator as the point of contact, and the delivery method). This fourth P is sometimes called Place, referring to the channel through which a product or service is distributed to the clients.

Although each funding agency will have its own (usually very specific) requirements, several elements of a proposal are fairly standard, and are often presented in the following order:

1. Title Page
2. Executive Summary
3. Introduction (Statement of the Problem, Purpose of Research or Goals, and Significance of Research)
4. Literature Review
5. Project Description or Program (Objective)
6. Project Narrative (Methods, Procedures, Outcomes or Deliverables, and Dissemination)
7. Project Evaluation
8. Personnel
9. Budget and Budget Justification
10. Timelines
11. Qualifications

Format the proposal so that it is easy to read. Use headings to divide the proposal into sections. If the proposal is long, include a table of contents with page numbers.

12.4.1   Title Page

The title page usually includes a brief (yet explicit) title for the research project, the names of the principle investigator(s), the corporate or institutional affiliation of the applicants (the department and university), name and address of the granting agency, project dates, amount of funding requested, and signatures of university personnel authorizing the proposal (when necessary). Most funding agencies have specific requirements for the title page; follow these requirements closely.

12.4.2   Executive Summary

The executive summary is a short, information-packed summary of the proposal. In one or two paragraphs, state the purpose of the proposal, the essentials of the program, and the total expense of the budget. This should not exceed one page. A reader should finish the summary knowing the basic information. Only an interested reader needs to read more. Write the executive summary after you have finished the rest of the report.

Executive Summary	Comments
(1) Real-time scheduling is a crucial factor in multiprocessor system and application performance. (2) The purpose of this project is to study the relative merits of the best fit and worst fit selection algorithms used in real-time scheduling. (3) The first goal of the project is to produce a memory management mechanism with the test sets and results for software developers using algorithms. (4) The second goal is to develop a very specific set of rules for soft error reduction. (5) The results will be valuable to software developers when choosing between the best fit and worst fit selection algorithms. (6) The project for attaining the results will be performed by a team of hardware and software experts for multiprocessor distributed systems. (7). The project will be completed in eight months and delivered in a technical report. (8) The cost for the eight-month period is $125,000.	Executive summary highlights the four Ps. The first sentence gives some context by defining the product. The project's goals are clearly stated in the second, third, and fourth sentences. The value of the project is outlined in sentence (5). The promotion of the project team is presented in sentence (6). Sentence (7) describes the placement of the project. Sentence (8) gives the pricing/cost.

12.4.3   Introduction

The introduction should cover the key elements of your proposal, including a statement of the problem, the purpose of research, research goals or objectives, and significance of the research. The statement of problem should provide the background and rationale for the project, and establish the need and relevance of the research. How is your project different from previous research on the same topic? Will you be using new methodologies or covering new theoretical territory? The research goals or objectives should identify the anticipated outcomes of the research and should match up to the needs identified in the statement of problem. List only the principal goal(s) or objective(s) of your research and save sub-objectives for the project narrative.

Introduction: One of the most important and challenging problems facing the designer of an autonomous robot is how the robot will determine its location. Recently, several algorithms have been suggested for successfully localizing a robot in a land-marked environment based on sensor readings. This project will focus on the less understood problem of collaborative multi-robot localization, that is, multiple robots working together to determine their position in the environment. Multiple robots working collaboratively allows the observations of one robot to aid others in their own localization, which can be very useful in situations, such as when not all robots have the same kind of sensor or when there are relatively few landmarks in the environment. The purpose of this project is to develop a method for multiple robots to cooperatively determine their positions and learn the localization reliabilities of fellow robots. These methods will be tested and developed in simulation on the computer for the remainder of the current fiscal year, and then will be tested on real robots. Because only one robot is currently available, this grant will be used to purchase another similar robot to simulate the collaborative multi-robot environment.

12.4.4   Literature Review

Many proposals require a literature review. Reviewers want to know whether you have done the necessary preliminary research to undertake your project. Literature reviews should be selective and critical, not exhaustive. Reviewers want to see your evaluation of previous research and development.

12.4.5   Project Description or Program (Objective)

State explicitly what you propose to do. Some descriptions also include a "scope" statement—an explicit statement of what you will not be doing to help limit the task. Explain your approach to the problem in detail. Explain your approach to solving the problem by answering the following questions:

- What are the technical specifications for the proposed project?
- How will current research, such as recent articles on the subject or other projects of a similar kind, be used to help solve the problem?
- How does your work fit into a larger project?

Included in your project description, you should have three subsections: (1) objectives, (2) methods, and (3) evaluation. You do not need to use these sections as subheadings, but you do need to clearly explain all three aspects of the project. As shown by the following example, your objectives must be tangible, specific, concrete, measurable, and achievable in a specified time period.

Objective: The last few years have witnessed an explosion in bioengineering digital signal processing of genomic sequence and protein structural data. These data contain a wealth of information about how organisms function and how they came to be. Currently, bioengineers have only a general description and understanding of how the Digital Signal Processing (DSP) algorithms work in deciding which algorithm is more appropriate for their application. I propose to quantify the performance of each DSP algorithm given varying sets of memory requests. This data will allow developers to compare the performance trade-off of each algorithm based on the expected memory request set for their application. This efficiency will cut the time for testing new applications by 54%. The first goal of the project is to produce a reference table with the test sets and results for bioengineers using DSP algorithms. The second goal is to develop a very specific set of rules for when to use each DSP algorithm. Whereas the first goal can be attained in the 10-week period (process), attaining the second goal will depend on the results of the data (product). The project for attaining the two goals will be completed in eight months. It will improve current ability of bioengineers and scientists and healthcare practitioners to access, manipulate, and interpret the rapidly accumulating mountain of data for medical and industrial applications.

12.4.6 Project Narrative

In your project narrative, you need to do the following:

- Describe the specific activities that will be implemented to accomplish your project objectives.
- Enable the reviewers to visualize the implementation of the project.
- Match the previously stated objectives.
- Provide the order and timing for the tasks.
- Defend your chosen methods, especially if they are new or unorthodox.

12.4.7 Project Evaluation

Building evaluation into a project is an important part of engineering design. You need to consider how you will evaluate whether the project is successful. How will you measure whether the project meets its goal? By including a mechanism for evaluation in your proposal, you indicate that achieving the objective is a serious goal. You also provide the best means for others to learn from your experience. Two types of formal evaluation are common:

1. Product measurement (e.g., test a computer program's performance under various conditions for versatility, accuracy, and speed)
2. Process Analysis (e.g., analyze the milestones, such as the ability of a prototype to integrate with other components of a project)

Project Narrative: The Research & Development (R&D) project will be at the forefront of theoretical and application-led research in fuzzy logic control. Initially the project will be focusing on using fuzzy logic control in multiprocessor real-time scheduling. This unique application will lead to some interesting results and a number of publications. An important strand of the R&D project will be collaborative work with Delft University of Technology in the Netherlands on the use of neural networks and fuzzy logic control in medical applications. The work will be concentrated on the role of neural networks in the classification of stress fractures of the tibia. Bringing team's expertise on fuzzy logic control to the problem, we will combine fuzzy logic and neural networks to tackle the same problem and perform an exhaustive investigation that will lead to a fault tolerant fuzzy logic control method. Other medical applications of the fuzzy logic control will be developed based on the modeling of nursing intuition and predicting pulmonary emboli.

Project Evaluation: Based on the call for proposals from the Chief Engineers Office, the project will be developed using the following evaluation criteria:

- Feasibility: Are the stated objectives logical, and will they lead toward proving the type, size, and location of the market opportunities? Does the research plan offer an original and innovative approach to the problem? Can the research plan reasonably be completed in the requested grant period?
- Importance of the Problem: Does the proposal provide sufficient justification for the importance of the problem and clearly indicate the anticipated commercial potential of the proposed research?
- Investigator and Resource Qualifications: Is adequate bibliographic information provided to document that the project director, other key staff, and any consultants have the appropriate training and experience to perform the proposed research plan?
- Budget: Is the budget appropriate for the proposed research plan, and is sufficient budget detail provided to indicate clearly how the funds would be used?
- Commercial Potential: Does the proposal provide sufficient explanation of the commercial potential for the project?

12.4.8   Personnel

Explain staffing requirements in detail, and make sure that staffing makes sense. Be very explicit about the skill sets of the personnel already in place (you will probably include their Curricula Vitae as part of the proposal). Explain the necessary skill sets and functions of personnel you will recruit. To minimize expenses, phase out personnel who are not relevant to later phases of a project.

Personnel			
Principal Investigator, 25%, Fiscal Year	$15,000	$0	$15,000
Project Associate, 10%	$0	$3,000	$3,000
Research Assistant, 50%	$9,000	$0	$9,000
Secretary, 50%	$7,000	$0	$7,000
Subtotal	$31,000	$3,000	$34,000
Staff Benefits (30% of S&W)	$9,300	$900	$10,200
Subtotal	$40,300	$3,900	$44,200
Consultants			
Paul Lee, $200/day, 2 days	$400	$0	$400

12.4.9   Budget and Budget Justification

The budget spells out project costs and usually consists of a spreadsheet or table with the budget detailed as line items and a budget narrative (also known as a budget justification) that explains the various expenses. Even when proposal guidelines do not specifically mention a narrative, be sure to include a one- or two- page explanation of the budget. Often, these two sections use a short paragraph or two to introduce graphic elements, such as Gantt Charts and Tables, to represent the proposed schedule. If necessary, rationale for the schedule or budget can also be presented in this section.

	Sponsor Agency	Company	Total
Equipment			
Optical methanometer	$5,000	$0	$5,000
Computer network system	$0	$3,900	$3,900
Materials and Supplies			
Glassware	$300	$0	$300
Chemicals	$200	$0	$200
Subtotal	$500	$0	$500
Travel			
Principal Investigator consultation with sponsor: Marion, IA, to Washington, D.C., and return. 1 person, 2 days			
Airfare	$800	$0	$800
Per Diem @ $100/day	$200	$0	$200
Local Transportation	$25	$0	$25
Subtotal	$1,025	$0	$1,025
Total Direct Costs	$6,525	$3,900	$10,425

12.4.10 Timelines

Explain the timeframe for the research project in some detail. When will you begin (and complete) each step? It may be helpful to reviewers if you present a visual version of your timeline. For less complicated research, a table summarizing the timeline for the project will help reviewers understand and evaluate the planning and feasibility.

Task	Months after Expected Start of 12/1/2008								
	1	2	3	4	5	6	7	8	9
Visit NanoTech	XXXXX								
Visit to Ames, IA Facility			XXXXXXXX						
Study Lathe Operation		XXXXX							
Define Weld Implementation		XXXXX							
Design Concepts for Fixture			XXX						
Design Selection for Fixture				XXXX					
Detail Design of Fixture					XXXX				
Fixture Fabrication							XXXX		
Welding Implemented							XXXXXXXXXX		
NanoTech Presentations		X			X			X	

12.4.11 Qualifications

This section presents another argument for why you should be allowed to undertake the project, usually by identifying professional and academic qualifications, experience, and attributes (less important) that make you (or your team) a suitable candidate for completing the plan.

12.5 A Brief Checklist before Submitting Your Proposal

Before submitting your grant proposal, check the following:

- Have you presented a compelling case?
- Have you made your hypotheses explicit?
- Does your project seem feasible? Is it overly ambitious? Does it have other weaknesses?
- Have you stated the means that grantors can use to evaluate the success of your project after you have executed it?

CURRICULUM VITAE

Paul Lee

Present Address

Department of Bioengineering

Telephone xxx-278-3140

MEMS, INC

5860 Ashwood Dr.

Marion, IA 52302

E-mail: john.lee@mems.com

Positions Held

(List only positions related to professional and academic achievements, and to your ability to conduct this research.)

Research Projects

Publications

Professional and Academic Presentations

Fellowships

Grants

Education

Honors and Awards

Scientific Societies (List the memberships you have in scientific societies.)

Languages (List any foreign languages that you can read, write or speak.)

Special Skills (List any special skills relevant to your ability to conduct research, such as scuba diver or computer skills.)

Teaching Experience

(You should fill in each of the above categories if appropriate. Do not include the category if it is not relevant to you.)

Bibliography

Alred, G. J., *Business Writer's Handbook*. 6th ed. New York: St. Martin's, 2000.

American Management Association, *The AMA Style Guide for Business Writing*. New York: AMACOM, 1996.

Beamer, L. and Varner, I. I., *Intercultural Communication in the Global Workplace*. 2nd ed. Boston: McGraw-Hill/Irwin, 2001.

Chaney, L. H. and Martin, J. S., *Intercultural Business Communication*. 2nd ed. Upper Saddle River, NJ: Prentice Hall, 2000.

Cialdini, R. B., *Influence: The Psychology of Persuasion*. Rev. ed. New York: Morrow, 1993.

Downey, R., Boland, S., and Walsh, P., *Communications Technology Guide for Business*. Boston: Artech House, 1998.

Eckhouse, B. E., *Competitive Communication: A Rhetoric for Modern Business*. Rev. ed. New York: Oxford University Press, 1999.

Fearn-Banks, K., *Crisis Communications: A Casebook Approach*. 2nd ed. Mahwah, NJ: Lawrence Erlbaum Associates, 2002.

Gardner, H., *Changing Minds: The Art and Science of Changing Our Own and Other People's Minds*. Boston, MA: Harvard Business School Press, 2004.

Geffner, A. B., *How to Write Better Business Letters*. 3rd ed. Hauppauge, N.Y: Barron's, 2000.

Geffner, A. B. and NetLibrary, Inc., *Business English: A Complete Guide to Developing an Effective Business Writing Style*. 3rd ed. Hauppauge, NY: Barron's Educational Series, 1998.

Guilar, J. D., *The Interpersonal Communication Skills Workshop: Listening, Assertiveness, Conflict Resolution, Collaboration*. New York: AMACOM, 2001.

Harvard Business Essentials: Business Communication. Boston: Harvard Business School Press, 2003. (The Harvard Business Essentials series)

Harvard Business Review on Effective Communication. Boston: Harvard Business School Press, 1999. (The Harvard Business Review paperback series)

Holtz, S., *Corporate Conversations: A Guide to Crafting Effective and Appropriate Internal Communications*. New York: AMACOM, 2004.

Jablin, F. M. and Putnam, L., *The New Handbook of Organizational Communication: Advances in Theory, Research, and Methods*. Thousand Oaks, CA: Sage Publications, 2001.

Levine, R. V., *The Power of Persuasion: How We're Bought and Sold*. Hoboken, NJ: John Wiley & Sons, 2003.

McLeary, J. W., *By the Numbers: Using Facts and Figures to Get Your Projects and Plans Approved*. New York: American Management Association, 2000.

Munter, M., *Guide to Managerial Communication: Effective Business Writing and Speaking*. 4th ed. Upper Saddle River, NJ: Prentice Hall, 1997.

New, C. C., *How to Write a Grant Proposal*. Hoboken, NJ: John Wiley, 2002.

Pan, Y., Scollon, S. B. K., and Scollon, R., *Professional Communication in International Settings*. Malden, MA: Blackwell Publishers, 2002.

Pearce, Terry. *Leading out Loud: Inspiring Change through Authentic Communication*. New and rev. ed. San Francisco: Jossey-Bass Publishers, 2003.

Rankin, E., *The Work of Writing: Insights and Strategies for Academics and Professionals*. San Francisco: Jossey-Bass Publishers, 2001.

Ryan, K., *Write up the Corporate Ladder: Successful Writers Reveal the Techniques That Help You Write with Ease and Get Ahead.* New York: American Management Association, 2003.

Simmons, J., *We, Me, Them, & It: The Powers of Words in Business.* New York and London: Texere, 2002.

Stockard, O., *The Write Approach: Techniques for Effective Business Writing.* San Diego: Academic Press, 1999.

Tingley, J C., *The Power of Indirect Influence.* New York: AMACOM, 2001.

Whalen, D. J., *I See What You Mean: Persuasive Business Communication.* Thousand Oaks, CA: Sage Publications, 1996.

Wiener, V., *Power Communications: Positioning Yourself for High Visibility.* New York: New York University Press, 1994.

Weissman, J., *Presenting to Win: Persuade Your Audience Every Time.* Upper Saddle River, NJ: Financial Times/Prentice Hall, 2003.

Worth, R., *Webster's New World Business Writing Handbook.* Indianapolis: Wiley Publishing, 2002.

13

How to Effectively Prepare Engineering Reports

Engineers write reports for many reasons, including the documentation of project progress and designs. As an engineer working on the design of an airplane seat, you might write the following two reports:

- Design report: Propose a new design for the seat, including documentation of tests performed to validate design.
- Progress report: Update the progress on the construction of a test seat, including prototyping and tooling evaluation.

Combining elements from these two reports helps to assess whether the new design should replace the existing design.

13.1 Writing an Effective Progress Report

Once you have written a successful proposal and have secured the resources to complete a project, you are expected to update the client on the progress of that project. This updating is usually handled by a progress report. Engineers write progress reports to keep interested parties informed about what has been done on a project and about what remains to be done. Often, the reader is the engineer's supervisor. Therefore, the tone should be serious and respectful.

Progress reports are common in engineering. As the name suggests, they document ongoing projects. They might be one-page memos or long, formal documents. Such a report is aimed at whoever assigned the project. Its goal is to enable the manager or sponsor of a project to make informed decisions about the future of the project. In the progress report, you explain any or all of the following:

- How much of the work is complete
- What part of the work is currently in progress
- What work remains to be done

- What problems or unexpected things, if any, have arisen
- How the project is going in general

13.1.1   Functions of Progress Reports

Although progress reports are often in the form of a memo, the writer should be careful to write formal, standard prose. Progress reports represent not only the writer's work, but also the writer's organizational and communication skills. Progress reports have several important functions:

- Reassure recipients that you are making progress, that the project is going smoothly, and that it will be complete by the expected date
- Provide recipients with a brief look at some of the findings or some of the work of the project
- Give recipients a chance to evaluate your work on the project and to request changes
- Give you a chance to discuss problems with the project and thus forewarn recipients
- Force you to establish a work schedule, so that you will complete the project on time

13.1.2   Format of Progress Reports

You write a progress report about the progress you have made on a project over a certain period of time. The project can be the design, construction, or repair of something, the study or research of a problem or question, or the gathering of information on a technical subject. You write progress reports when it takes well over three or four months to complete a project. Progress reports can be structured in several ways. The following suggested pattern helps the writer cover essential material.

13.1.2.1   Heading

If the progress report is a memo, it should contain the following standard elements:

- Date: Date the memo is sent
- To: Name and position of the reader
- From: Name and position of the writer
- Subject: A clear phrase that focuses the reader's attention on the subject of the memo

---

MEMORANDUM

Date: June 15, 1998

To: Bob Lafferty, Plant Manager

From: John Hill, Manufacturing Engineer

Subject: Progress report on quartz etch rate project

13.1.2.2   Purpose Statement

Because the reader is busy, get right to the point. Imagine you are meeting the reader in the hall, and you say, "I wanted to talk to you about this." Use the same strategy for the first line of the memo's body. Try saying out loud, "I wanted to tell you that" and then start writing what ever comes after that prompt. Often, such a sentence will begin something like this: "Progress on setting up the new program in testing is going very well." If there is a request somewhere in the memo, make it explicit up front; otherwise, your reader may miss it.

Bob, here is the update on the quartz etch rate project that you requested. I have included some general overview of the project, a review of its subject and scope along with details on the progress we have made on each objective in the project. I conclude with a tentative outline of the report.

Purpose. The purpose of the project is to obtain quartz etch rate data for future reference. Instructions will be provided that will ensure accurate results when followed.

13.1.2.3   Background

In many instances, the progress report's reader (an engineering manager, for instance) is responsible for several projects. Therefore, the reader expects to be oriented as to what your project is, what its objectives are, and what the status of the project was at the time of the last reporting.

Usually, in the same paragraph as the purpose statement, the writer gives the reader some background information. If the occasion demands a written progress report instead of a quick oral report, it is probably the case that the reader needs to be reminded of the details. Tell the reader what the project is, and clarify its purpose and time frame for completion. If earlier progress reports have been submitted, you might make a brief reference to them.

Background: The project will develop a method for obtaining quartz etch rates from the Polyflow vertical quartzware cleaner. A report on the findings will be delivered to you and the diffusion engineering staff. A procedure for collecting this data in the future will be created.

13.1.2.4   Work Completed

The next section of a progress report explains what work has been completed during the reporting period. Specify the dates of the reporting period, and use active voice verbs to give the impression that you, or you and your team, have been busy. You might arrange this section chronologically (following the actual sequence of the tasks being completed), or you might divide this section into subparts of the larger project and report on each subpart in sequence. Whatever pattern you use, be consistent.

Work Completed: Execution of the etch rate tests has been flawless. A method for measuring the disks for thickness and surface roughness was adopted. This method allows for some disks to be in all phases of the testing process at once (pre-measurement for thickness, pre-measurement for surface roughness, etching, post-measurement for thickness and post-measurement for surface roughness). Detailed instructions were created, which allow anyone to repeat a procedure. Several operators were asked to perform an etch test using only the written instructions and no further information. They were able to repeat the test without error.

13.1.2.5   Problems

If the reader is likely to be interested in the glitches you have encountered along the way, mention the problems you have encountered and explain how you have solved them. If you have not yet been able to solve certain problems, explain your strategy for solving them and tell the reader when you think you will have them solved.

Snags are expected in a project. Do not hide from them; explain what they are and how they might affect key areas of the job, such as timing, price, or quality. If the problem occurred in the past, you can explain how you overcame it. This is least serious; in fact, you look good. If the problem occurs again (now or in the future), explain how you hope to overcome it, if you can.

> **Problem:** The main problem is to collect enough tests under all the critical test conditions. I plan to implement a data acquisition system for rapid data collection.

Progress reports are not necessarily for the benefit of only your manager, who is the reader of these reports. Often, you benefit from the reporting because you can share or warn your manager about problems that have arisen. In some situations, your manager might be able to direct you toward possible solutions. In other situations, you might negotiate a revision of the original objectives, as presented in the proposal.

13.1.2.6 Work Scheduled

In this section, you discuss your plan for meeting the objectives of the project. Here, you specify the dates of the next segment of time in the project and outline a schedule of the work you expect to accomplish during the period. Arranging this section by dates that represent deadlines is a good idea. To finish the progress report, you might add a sentence that evaluates your progress thus far.

> **Work Scheduled:** 31 of the 39 necessary etches have been performed. The remaining 8 tests will be completed by the end of the month. Write, and submit for approval, project report. Add etch rate procedure to the Polyflow operation specification.

13.1.2.7 Status Assessment

Here, you assess whether you will meet the objectives in the proposed schedule and budget. In many situations, this section is the bottom line for the reader. In some situations, such as the construction of a highway, failure to meet the objectives in the proposed schedule and budget can result in the engineers having to forfeit the contract. In other situations, such as a research project, the reader expects that the objectives will change somewhat during the project.

> **Status Assessment:** This progress report has updated you on the status of the quartz etch rate project. As stated, I am on schedule and should complete the project by the original deadline, December 6, 2008. Because preliminary review has raised interesting questions about the test procedures, I request permission to modify my original objectives, discussed in the proposal, to focus on those test procedures. In doing so, I believe that I will attain reliability and productivity for the quartz etch rate.

13.1.3   Checklist for Progress Reports

To report this information, you combine two of the following three organizational strategies: time periods, project tasks, or report topics. As you reread and revise your progress report, watch out for problems such as the following:

- Make sure you use the right format. Remember, the memo format is for internal progress reports; the business-letter format is for progress reports written from one external organization to another. (Whether you use a cover memo or cover letter is your choice.)
- Write a good introduction. In the introduction, state that this is a progress report, and provide an overview of the contents of the progress report.
- Make sure to include a description of the final report project.
- Use one or a combination of the organizational patterns in the discussion of your work on the final report.
- Use headings to mark off the different parts of your progress report, particularly the different parts of your summary of work completed on the project.
- Use lists as appropriate.
- Provide specifics. In the report, avoid relying on vague, overly general statements about the work you have completed.
- Be sure to address the progress report to the real or realistic audience—the real interested party.

Assume there will be a nonspecialist reading your progress report. However, do not avoid discussion of technical aspects of the project—just use language that nonspecialists can understand.

13.2   Develop Informative Design Reports

Design reports are written to introduce and document engineering designs. In general, design reports have two audiences: (1) other engineers interested in how the design works and how effective the design is, (2) management interested in the application and effectiveness of the design. Design reports are often organized as follows:

- Summary
- Introduction
- Discussion
- Conclusions
- Appendices

13.2.1 How to Write a Summary

The summary, sometimes labeled the abstract or executive summary, is a concise synopsis of the actual design, the motivation for having the design, and the design's effectiveness. The author should assume that the reader has some knowledge of the subject, but has not read the report. For that reason, the summary should provide enough background that it can stand alone. Note that if the summary is called an abstract, you are usually expected to target a technical audience in the summary. Likewise, if an executive summary is requested, you should target a management audience in the summary.

Summary

Electro-dynamic actuators are based on Lorentz force effects. The current-carrying coil is stationary in the former case, whereas in the latter it is moving. These are ideally suited when large currents are possible, even with lower voltages. This design report presents a lightweight electro-dynamic actuator that has very high reliability and efficiency. An integrated velocity sensor is implemented for velocity feedback control. The new actuator arrangement provides a collocated force actuator and velocity sensor device so that, in principle, an unconditionally stable direct velocity feedback loop could be implemented. Self-sensing active vibration damping is provided if external sensors cannot be collocated with the actuators or if these sensors add too much weight or cost

13.2.2 How to Write an "Introduction"

The "Introduction" of a design report identifies the design problem, the objectives of the design, the assumptions for the design, the design alternatives, and the selection of the design being reported. Also included for transition is a mapping of the entire report. Note that in longer reports, the selection of design is often a separate section.

> **Introduction**
>
> Electro-dynamic actuators are the final elements in a control system. They receive a low-power command signal and energy input to amplify the command signal, as appropriate, to produce the required output. Applications range from simple low-power switches to high-power hydraulic devices, which operate flaps and control surfaces on aircraft, valves, and car steering. This design report presents a lightweight electro-dynamic actuator that has superior reliability and efficiency.

13.2.3 How to Present Your Discussion

The discussion presents the actual design, the theory behind the design, the problems encountered (or anticipated) in producing the design, how those problems were (or could be) overcome, and the results of any tests on the design. Note that this part usually consists of two, three, or four main headings. Regarding the actual names of these headings, pay close attention to what your client/customer requests. Also consider what would be a logical division of electrical, mechanical, and thermal aspects for your particular design.

Discussion

TWC Actuators are electro-dynamic actuators. Actuators are grouped as follows:

Electromechanical

- Electromagnetic—exploits the mutual attraction of soft ferrous materials in a magnetic field. The device has one coil that provides the field energy and the energy to be transformed. The attractive force is unidirectional so a return device of some type is needed, often a spring. Relays or solenoids based on this principle are widely used in cars to switch a range of electrical equipment with a current demand of more than about 10 amps—fans, headlights, horn, wipers.
- Electro-dynamic—based on the (Lorentz) force generated when a current carrying conductor (often in the form of a coil) is held in a magnetic field. DC motors are frequently used as part of an actuator system.

Fluid mechanical

- Pneumatic—A common device is the pneumatic cylinder.
- Hydraulic—A common device is the hydraulic cylinder.

Significant trend is the move away from hydraulic to electrical devices. This is driven partly by the desire to have cleaner systems (no hydraulic fluid) and making integration with other (normally electrical) control systems easier to achieve. New cars are often now fitted with electric power assisted steering instead of the hydraulic system that was the only system available until recently. Developments are progressing with electrical assistance of car braking systems. This trend to electrical systems is also present in the aviation industry, but the very high-power densities and forces required from some actuators mean that this would be more difficult.

Based on a lumped parameter model, important aspects of the design of an electro-dynamic, inertial actuator with internal velocity sensor are found. First the electromagnetic design procedure that optimizes the actuation force calculated with a Finite Element Analysis (FEA) at a given weight by balancing the amount of wire, and the strength of the magnetic field in the air gap is presented. It turns out that a rather small magnetic field in the air gap is sufficient to generate a maximum force at a given mass and power input, as long as sufficient wire fits into the air gap and the strongest available permanent magnets are used.

Using self-sensing in an electro-dynamic actuator for broadband active vibration damping requires compensation of the actuator resistance and of the self-inductance of the actuator with an appropriate shunted circuit. To reduce power consumption the actuator resistance should be small, but for robustness of self-sensing and a large bandwidth, a large resistance is required. A high transducer coefficient is important to get high sensitivity of the induced voltage that is proportional to the vibration velocity of an attached mechanical structure. However, a large transducer coefficient implies a strong magnetic field that also increases the self-inductance so that the measurement bandwidth potentially is reduced. In this study, to eliminate the first trade-off between power consumption and robustness, an actuator with a primary driving coil and a secondary measurement coil is proposed. The primary coil is optimized for driving by choosing a small resistance, whereas the secondary coil is optimized for sensing by choosing a large resistance. Transformer coupling occurs between the two coils. Decreasing the cross section of the secondary coil could reduce the coupling. However, there is a geometric limit on the reduction of the cross section of the secondary coil. Because of this limitation, an analogue electronic compensation scheme is proposed to compensate for the transformer coupling between the primary and the secondary coil. Feedback of the sensed velocity in the secondary coil is implemented, and the experimental vibration damping that results at a plate is presented. Results are compared to self-sensing vibration damping, active vibration damping using a velocity sensor, and passive damping by means of the same weight as the actuator.

13.2.4 How to Deliver Your Conclusion

The "Conclusions" section summarizes the design and testing work completed, and assesses how well the design meets the objectives presented in the "Introduction." Note that if the design does not meet the objectives, you should analyze why the design did not succeed and what could be modified to make the design a success. Besides summarizing the work and analyzing whether the objectives were met, the "Conclusions" section also gives a perspective for how the design will be used in the future.

Conclusions

Self-sensing active vibration damping is advantageous if external sensors cannot be collocated with the actuators or if these sensors add too much weight or cost. When self-sensing, electro-dynamic actuators are used, damping is directly added to an attached structure without the need of potentially destabilizing electronic integrators or differentiators. In this paper, self-sensing control with a shunted resistor, positive current feedback, and induced voltage feedback are investigated in simulations and experiments. Experiments with a commercial shaker on a clamped plate demonstrate that its vibration attenuation is increased from 5 to 15 dB, and the control bandwidth is more than doubled when the appropriate control scheme is used.

13.3   Summary

As a practicing engineer, you will need to write reports to convey your ideas to managers, other engineers, and customers. The final report of any project is not just a formality. It is a primary product of the effort and is often the basis for the evaluation of the reporter's professional abilities. The report is also a service to those in need of the information.

A good engineering report should promote readability and reflect the scientific method of attack, which proceeds with an objective, method, results, and conclusions. Reporting on a project in the sequence in which it is done is logical, and many engineering reports are organized on this basis. Two improvements to the logical sequence are the addition of an abstract or executive summary and the insertion of headlines. These two features facilitate "scanning" of the report. Thus, a busy executive or engineer may quickly assess the major findings and conclusions of the report and easily find further details as required.

Bibliography

Alred, G. J., *Business Writer's Handbook*. 6th ed. New York: St. Martin's, 2000.

American Management Association, *The AMA Style Guide for Business Writing*. New York: AMACOM, 1996.

Beamer, L. and Varner, I. I., *Intercultural Communication in the Global Workplace*. 2nd ed. Boston: McGraw-Hill/Irwin, 2001.

Chaney, L. H. and Martin, J. S., *Intercultural Business Communication*. 2nd ed. Upper Saddle River, NJ: Prentice Hall, 2000.

Cialdini, R. B., *Influence: The Psychology of Persuasion*. Rev. ed. New York: Morrow, 1993.

Downey, R., Boland, S., and Walsh, P., *Communications Technology Guide for Business*. Boston: Artech House, 1998.

Eckhouse, B. E., *Competitive Communication: A Rhetoric for Modern Business*. Rev. ed. New York: Oxford University Press, 1999.

Fearn-Banks, K., *Crisis Communications: A Casebook Approach*. 2nd ed. Mahwah, NJ: Lawrence Erlbaum Associates, 2002.

Gardner, H., *Changing Minds: The Art and Science of Changing Our Own and Other People's Minds*. Boston, MA: Harvard Business School Press, 2004.

Geffner, A. B., *How to Write Better Business Letters*. 3rd ed. Hauppauge, N.Y: Barron's, 2000.

Geffner, A. B. and NetLibrary, Inc., *Business English: A Complete Guide to Developing an Effective Business Writing Style*. 3rd ed. Hauppauge, NY: Barron's Educational Series, 1998.

Guilar, J. D., *The Interpersonal Communication Skills Workshop: Listening, Assertiveness, Conflict Resolution, Collaboration*. New York: AMACOM, 2001.

Harvard Business Essentials: Business Communication. Boston: Harvard Business School Press, 2003. (The Harvard Business Essentials series)

Harvard Business Review on Effective Communication. Boston: Harvard Business School Press, 1999. (The Harvard Business Review paperback series)

Holtz, S., *Corporate Conversations: A Guide to Crafting Effective and Appropriate Internal Communications*. New York: AMACOM, 2004.

Jablin, F. M. and Putnam, L., *The New Handbook of Organizational Communication: Advances in Theory, Research, and Methods*. Thousand Oaks, CA: Sage Publications, 2001.

Levine, R. V., *The Power of Persuasion: How We're Bought and Sold*. Hoboken, NJ: John Wiley & Sons, 2003.

McLeary, J. W., *By the Numbers: Using Facts and Figures to Get Your Projects and Plans Approved*. New York: American Management Association, 2000.

Munter, M., *Guide to Managerial Communication: Effective Business Writing and Speaking*. 4th ed. Upper Saddle River, NJ: Prentice Hall, 1997.

New, C. C., *How to Write a Grant Proposal*. Hoboken, NJ: John Wiley, 2002.

Pan, Y., Scollon, S. B. K., and Scollon, R., *Professional Communication in International Settings*. Malden, MA: Blackwell Publishers, 2002.

Pearce, Terry. *Leading out Loud: Inspiring Change through Authentic Communication*. New and rev. ed. San Francisco: Jossey-Bass Publishers, 2003.

Rankin, E., *The Work of Writing: Insights and Strategies for Academics and Professionals*. San Francisco: Jossey-Bass Publishers, 2001.

Ryan, K., *Write up the Corporate Ladder: Successful Writers Reveal the Techniques That Help You Write with Ease and Get Ahead*. New York: American Management Association, 2003.

Simmons, J., *We, Me, Them, & It: The Powers of Words in Business*. New York and London: Texere, 2002.

Stockard, O., *The Write Approach: Techniques for Effective Business Writing*. San Diego: Academic Press, 1999.

Tingley, J C., *The Power of Indirect Influence*. New York: AMACOM, 2001.

Whalen, D. J., *I See What You Mean: Persuasive Business Communication*. Thousand Oaks, CA: Sage Publications, 1996.

Wiener, V., *Power Communications: Positioning Yourself for High Visibility*. New York: New York University Press, 1994.

Weissman, J., *Presenting to Win: Persuade Your Audience Every Time*. Upper Saddle River, NJ: Financial Times/Prentice Hall, 2003.

Worth, R., *Webster's New World Business Writing Handbook*. Indianapolis: Wiley Publishing, 2002.

14

Listening—Interactive Communication about Engineering Risk

14.1 Listening—A Forgotten Risk Communication Skill

Risks also accompany the benefits we enjoy from many engineering projects and technologies. Human activities and natural disasters could also pose risks to buildings, infrastructure, and other engineering establishments. Figure 14.1 presents some of the spectacular engineering failures or natural disasters that occurred in August 2007. Making wise choices requires understanding risks and benefits. Risk communication is a tool for creating an understanding of risks and benefits of engineering endeavors, closing the gap between lay people and experts, and helping people make more informed and healthier choices.

We often tell our customers, peers, and suppliers how to assess and manage risk. Engineering risks include technical risk, schedule risk, and cost risk. Many technical books about risk assessment and management have been published. However, risk management involves risk communication, which involves listening to identify various risks correctly.

In risk communication, listening means gathering and making sense out of information about technical risk, schedule risk, and cost risk. You cannot provide an engineering solution unless you understand your customer's problem; and you cannot gain team consensus unless you understand each team member's perceptions about the project merits, schedule, and cost. In all these cases, you must listen to others.

Risk communication theory and practice may include public participation and conflict resolution. An engineer can improve his or her level of risk communication by thinking strategically about listening and practicing effective listening skills. Your skill as a listener can make or break your success in risk management, project leadership, cross-functional teams, customer relationships, and business negotiations. Listening is the only way to learn how to improve:

- Your risk management, cost reduction, project leadership, engineering communication, and business negotiating skills
- Your ability to motivate customers and teams
- Your overall performance in business

- August 1, Minneapolis, Minn.: An eight-lane interstate bridge packed with cars broke into sections and collapsed into the Mississippi River, killing at least nine and injuring at least 60. The bridge was in the midst of repairs when it buckled and broke apart.

- August 1, Benaleka, Congo: A passenger train running between Ilebo and Kananga derailed after the brakes failed, killing about 100 people.

- August 14, Hunan province, China: A bridge undergoing construction collapsed in southern China, killing at least 28 people.

- August 25–27, Greece: Over 220 separate fires ravage the Greek countryside and endanger ancient Olynpic sites around Athens. At least 64 people die in the blazes.

FIGURE 14.1
Risk Spectrum: some of the disasters that occurred in August 2007.

The more effectively you listen, the more effectively you communicate about engineering risk and cost. However, only about 25% of listeners grasp the central ideas in business communications. To improve listening skills, consider the following:

Poor Listener	Effective Listener
Tends to "wool-gather" with slow speakers	Thinks and mentally summarizes, weighs the evidence, listens between the lines to tones of voice and evidence
Subject is dry, so tunes out speaker	Finds "what's in it for me"
Distracted easily; lets mind wander if thinks he or she knows what the person is going to say next	Fights distractions, sees past bad communication habits, knows how to concentrate
Takes intensive notes, but the more notes taken, the less value; has only one way to take notes	Has two to three ways to take notes and organize important information
Is over-stimulated, and tends to seek and enter into arguments	Does not judge until comprehension is complete

Inexperienced in listening to difficult material; has usually sought light, recreational materials	Uses "heavier" materials to regularly exercise the mind
Lets deaf spots or blind words catch his or her attention	Interprets color words and does not get hung up on them
Shows no energy output	Holds eye contact and helps speaker along by showing an active body state
Judges delivery—tunes out	Judges content, skips over delivery errors
Listens for facts	Listens for main ideas.
Interrupts the speaker	Lets the speaker finish before he or she begins to talk
Busy thinking about what he or she wants to say next	Lets himself or herself finish listening before he or she begins to speak

14.2   Listening—Harder Than Speaking and Writing

14.2.1   What Causes Listening to Fail?

The task of informing the public about various engineering risks is fraught with many problems. It is essential to overcome these problems if risk communication is to be improved. Failing to understand people's risk perceptions presents a major hazard in risk and cost communication. An effective risk and cost communication begins with listening to the people's voices about perceived risk and cost.

One of the best ways to begin to improve your listening skills is to have a better understanding of some of the most common behaviors you and others demonstrate when not listening effectively. Keep in mind that the following listening blocks should not always be considered bad. In certain situations, they can be effective at helping an individual achieve a particular result. The key to their effectiveness is to be aware of when and why you are using them.

Rehearsing—All your attention is on designing and preparing your next comment. You look interested, but your mind is going a mile a minute because you are thinking about what to say next. Some people rehearse entire chains of responses: "I'll say, then he'll say, and so on."

Judging—Negatively labeling people can be extremely limiting. If you prejudge somebody as incompetent or uninformed, you do not pay much attention to what that person says. A basic rule of listening is that judgments should only be made after you have heard and evaluated the content of the message.

Identifying—In this this block, you take everything people tell you and refer it back to your own experience, often launching into your story before they can finish theirs.

Advising—You are the great problem solver. You do not have to hear more than a few sentences before you begin searching for the right advice. However, while you are coming up with suggestions and convincing someone to just try it, you may miss what is most important.

Sparring—This block has you arguing and debating with people who never feel heard because you are so quick to disagree. In fact, your main focus is on finding things with which you disagree.

Being Right—Being right means you will go to great lengths (e.g., twist the facts, start shouting, make excuses or accusations, call up past sins) to avoid being wrong. You cannot listen to criticism, you cannot be corrected, and you cannot take suggestions for change.

Derailing—This listening block involves suddenly changing the subject. You derail the train of conversation when you get uncomfortable or bored with a topic. Another way of derailing is by joking.

Placating—Right … Absolutely … I know … Of course you are … Incredible … Really? You want to be nice, pleasant, and supportive. You want people to like you, so you agree with everything. You may half-listen just enough to get the drift, but you are not really involved.

Dreaming—When we dream, we pretend to listen but really tune the other person out while we drift about in our interior fantasies. Instead of disciplining ourselves to truly concentrate on the input, we change the channel to a more entertaining subject.

14.3　How to Listen to Voices of Customers about Risk

Risk communication is a complex, multidisciplinary, multidimensional, and evolving process of increasing importance in protecting customers and the public. Engineers use risk communication to give customers necessary and appropriate information and to involve them in making decisions that affect them, such as where to build waste disposal facilities.

The National Research Council (NRC) defines risk communication as "an interactive process of exchange of information and opinion among individuals, groups, and institutions." The definition includes "discussion about risk types and levels and about methods for managing risks." Listening to the voice of customers and the public is central to effective communication about topics of high concern:

- Give 100% Attention. Prove you care by suspending all other activities. Give your full attention to the person who is speaking. Make sure your mind is focused, too. Move your mind to concentrate on what the speaker is saying. You cannot fully hear this person's point of view or process the information when you argue mentally or judge what the person is saying before he or she has completed the thought. An open mind is a mind that is receiving and listening to information. If you feel your mind wandering, change the position of your body and try to concentrate on the speaker's words.

- Let the speaker finish before you begin to talk. Speakers appreciate having the chance to say everything they would like to say without being interrupted. When you interrupt, it looks like you are not listening, even if you really are.

- Let yourself finish listening before you begin to speak! You cannot really listen if you are busy thinking about what you want say next.

- Listen for main ideas. The main ideas are the most important points the speaker wants to emphasize. These points may be mentioned at the beginning or end of a talk, and repeated a number of times. Pay special attention to statements that begin with phrases such as "My point is … " or "The thing to remember is … ."

- Ask questions. If you are not sure you understand what the speaker has said, just ask. It is a good idea to repeat in your own words what the speaker said so that you can be sure your understanding is correct. For example, you might say, "When you said that no two zebras are alike, did you mean that the stripes are different on each one?"

- Give feedback. Sit up straight and look directly at the speaker. Now and then, nod to show that you understand. At appropriate points, you may also smile, frown, laugh, or be silent. These are all ways to let the speaker know that you are really listening. Remember, you listen with your face as well as your ears.

- Thinking fast. Remember, time is on your side. Thoughts move about four times faster than speech. With practice, you will find that while you are listening, you will also be able to think about what you are hearing, really understand it, and give feedback to the speaker.

- Respond. Responses can be both verbal and nonverbal (i.e., nods, expressing interest), but must prove you received the message, and more important, prove it had an impact on you. Speak at approximately the same energy level as the other person—then that person will know they really got through and will not have to keep repeating the message.

- Prove understanding. To say, "I understand" is not enough. People need some sort of evidence or proof of understanding. Prove your understanding by occasionally restating the gist of their ideas or by asking a question that proves you know the main idea. The important point is not to repeat what they have said to prove you were listening, but to prove you understand. The difference in these two intentions transmits remarkably different messages when you are communicating.

- Prove respect. Prove you take other views seriously. It seldom helps to say, "I appreciate your position" or "I know how you feel." You have to prove it by your willingness to communicate with others at their level of understanding and attitude. We do this naturally by adjusting our tone of vice, rate of speech, and choice of words to show that we are trying to imagine being where they are at the moment.

Listening to and acknowledging other people may seem deceptively simple, but doing it well, particularly when disagreements arise, takes true talent. As with any skill, listening well takes plenty of practice.

14.4   Listen Attentively: Understanding What Drives Perceived Risk

Attentive listening means thinking and acting in ways that connect you with the speaker. Although active listening usually happens naturally when we are very interested in what someone is saying, we can also choose to listen actively whenever we want to maximize the quality of our listening, both in terms of the effect it has on us and the effect it has on those to whom we are speaking. By contrast, when people "multitask" while someone speaks, they rarely listen effectively.

Of course, sometimes you will choose *not* to listen attentively. If listening is important to you, you may choose to reschedule that conversation. Otherwise you will multitask (i.e., letting your mind wander to think about something else, reading e-mail, doodling, and so forth).

You can take several simple steps to improve your listening skills. The quality of information exchanged, your personal experience as a listener, the experience of the person to whom you are listening, and your relationship with the listener will all benefit. The steps are:

- Get over yourself; give the speaker a solo.
- Stop multitasking.
- Recap regularly.
- Use connecting words.
- Use body language.

14.4.1 Get over Yourself; Give the Speaker a Solo

If you assert your own position at every opening in a conversation, you will eliminate many of the potential benefits of listening. In particular, the speakers will not feel respected by you, their thinking and brainstorming will be inhibited, and they may even withhold important information out of caution—or out of anger.

Wait until they finish making their points before you speak. Do not interrupt, even to agree with them, and do not jump in with your own suggestions before they explain what they have already done, plan to do, or have thought about doing. This includes being aware enough to stop yourself from doing any of the following:

- Making critical or judgmental faces or sounds
- Trying to "fix" their problems with quick suggestions
- Interrogating them to make them answer questions you have about their situations
- Trying to cheer them up or tell them things are not so bad
- Criticizing them for getting into their situations
- Telling them what you would do or have done in the past

All these responses interrupt what they are saying or change the direction of the conversation before they have an opportunity to get to their points. The first thing people bring up when they have something to say often *is not* the central point they will eventually make, whether they know it or not. Listening carefully for a while first gives the speaker and the listener both a chance to develop an understanding about what is exactly the issue.

Just to be sure the speaker has reached a stopping point and is ready to give up the floor, you can ask, for instance, "May I make a suggestion?" before you begin talking.

14.4.2 Stop Multitasking

Do not multitask if you are supposed to be listening. You wind up either listening to only part of what someone says or pretending to listen while you think about something else. You also sacrifice important nonverbal cues and information about the speaker's intent, confidence level, and commitment level. Even if you think that you can get enough of what the speaker is saying, while you multitask, to serve your immediate purposes, you should assume, as a general rule, that a person notices when you do not listen to him or her attentively.

If you are tempted to split your attention between listening and something else, ask yourself whether you can risk appearing disinterested, not to mention the negative impression that it is likely to make on the speaker.

Avoid allowing interruptions that cause you to lose concentration or split your attention. Eliminate background noise, ringing telephones, and impromptu visits by other people. Do not read e-mail, use a computer, or read something while someone else is talking to you.

If you find your attention wandering, use this trick: Decide why you do not want to listen. Think about what you might get out of listening, then choose whether to listen or not.

14.4.3   Recap Regularly

Very skilled listeners practice and become good at recapping both the facts and the level of importance (i.e., the emotional drift of the speaker) in a few brief words.

> Example: Amir tells Brenda that the company has introduced object-oriented (OO) technology into its organization by selecting a well-defined project "X" with hard schedule constraints to pilot the use of the technology. "Although many 'X' project personnel were familiar with the OO concept, it had not been part of their development process, and they have had very little experience and training in the technology's application," Amir said. "It is taking project personnel longer than expected to climb the learning curve. Some personnel are concerned, for example, that the modules implemented to date might be too inefficient to satisfy project 'X' performance requirements." After he reaches a stopping point, Brenda recaps, "So the risk is: Given the lack of OO technology experience and training, the possibility exists that the product will not meet performance or functionality requirements within the defined schedule. So, you're pretty worried about their indecision at this point, right?"

Metaphors are a compelling way to sum up what someone has been saying. For example, if someone describes how a project he or she was working on had to be done over and over again because of a glitch, you could say, "You're that guy who has to keep pushing a boulder up a hill, even though it always rolls back down again" [Sisyphus, from Greek mythology].

If you do not understand or are not sure about a point the person is trying to make, repeat a very brief portion of the part you did not understand and ask him or her to tell you more about it to help improve your understanding.

14.4.4   Use Connecting Words

Where it helps, use words that show you are connecting with what the person is saying, such as "uh huh," "OK," "yeah," "I get it," and so forth.

14.4.5 Use Body Language

Use positive body language, such as making frequent eye contact and facing the speaker squarely. Avoid negative body language such as frowning and looking away.

A great deal of research has been done about body language. Books have been written about it, and some people claim to be experts at interpreting it. For the rest of us, however, it is enough to be aware that body language exists, and we should use it constructively when we can.

The following example demonstrates all of the attentive listening techniques described in this section.

> Example: Carl steps into Wanda's office, frowning and looking at the floor, and asks her if she has a couple of minutes for an important problem. Wanda decides that Carl has important information to give her, and needs to feel better and revive his motivation by talking about his problem. Wanda decides to listen to Carl attentively, and asks Carl to close her door, which signals to others that she is unavailable, and turns down the volume on her computer, which mutes the music she was playing as well as the sound of her incoming e-mails. Wanda hits the "do not disturb" button on her phone, turns her chair to face Carl, and begins making and holding eye contact with him.
>
> Carl describes how a sudden rise in customer complaints had been traced to a previously undiscovered bug in the programming for a product delivered long ago. The problem is compounded by the fact that none of the people who originally worked on the programming are still with the company, adding considerably to the difficulty and the degree of anxiety being experienced by the team rushing to correct the problem. Wanda listens, without interrupting, occasionally saying "Uh huh" and "OK," while trying hard not to look angry or alarmed as the story deepens. Now and again, she repeats something Carl has just said, and asks him to elaborate on a particular point. When he appears to have told his entire story, she sums up with a metaphor: "To coin a phrase, Carl, after 30 straight days of perfect weather, everybody forgot his or her umbrellas, so now we're getting drenched. Is that about right?"

Engineering involves risk and unexpected cost. Risk communication represents the process of working with customers and the public to weigh the odds. The benefits of risk communication include improved decision making, both individually and collectively. The purpose of the exchange and the nature of the information have an impact on the benefits. Effective risk communication starts with listening to the people's specific concerns.

14.5   Thirteen Questions about Risk Communication

Risk is a combination of the probability of an event (usually adverse) and the nature and severity of the event. The main goal in understanding and communicating risk is to identify and impose priorities, and take appropriate actions to minimize risks. Listen to people's specific concerns. People often care as much about credibility, competence, and empathy as they do about risk levels, statistics, and details. Ask yourself and your project team the following 13 questions:

1. Why are we communicating?
2. Who is our audience?
3. What do our audiences want to know?
4. What issues or points do we want to emphasize?
5. How will we communicate?
6. How will we listen?
7. How will we respond?
8. Who will complete the plans? When?
9. What problems or barriers have we anticipated?
10. What are the opportunities for effective communications, and how can we maximize these opportunities?
11. What questions can we anticipate from the public in these risk situations?
12. What are the news media's responsibilities, and how can we help reporters meet them?
13. Have we succeeded in communicating all the risks and benefits of the engineering endeavors?

Bibliography

Alred, G. J., *Business Writer's Handbook*. 6th ed. New York: St. Martin's, 2000.

American Management Association, *The AMA Style Guide for Business Writing*. New York: AMACOM, 1996.

Beamer, L. and Varner, I. I., *Intercultural Communication in the Global Workplace*. 2nd ed. Boston: McGraw-Hill/Irwin, 2001.

Chaney, L. H. and Martin, J. S., *Intercultural Business Communication*. 2nd ed. Upper Saddle River, NJ: Prentice Hall, 2000.

Cialdini, R. B., *Influence: The Psychology of Persuasion*. Rev. ed. New York: Morrow, 1993.

Downey, R., Boland, S., and Walsh, P., *Communications Technology Guide for Business*. Boston: Artech House, 1998.

Eckhouse, B. E., *Competitive Communication: A Rhetoric for Modern Business*. Rev. ed. New York: Oxford University Press, 1999.

Fearn-Banks, K., *Crisis Communications: A Casebook Approach*. 2nd ed. Mahwah, NJ: Lawrence Erlbaum Associates, 2002.

Gardner, H., *Changing Minds: The Art and Science of Changing Our Own and Other People's Minds*. Boston, MA: Harvard Business School Press, 2004.

Geffner, A. B., *How to Write Better Business Letters*. 3rd ed. Hauppauge, N.Y: Barron's, 2000.

Geffner, A. B. and NetLibrary, Inc., *Business English: A Complete Guide to Developing an Effective Business Writing Style*. 3rd ed. Hauppauge, NY: Barron's Educational Series, 1998.

Guilar, J. D., *The Interpersonal Communication Skills Workshop: Listening, Assertiveness, Conflict Resolution, Collaboration*. New York: AMACOM, 2001.

Harvard Business Essentials: Business Communication. Boston: Harvard Business School Press, 2003. (The Harvard Business Essentials Series)

Harvard Business Review on Effective Communication. Boston: Harvard Business School Press, 1999. (The Harvard Business Review Paperback Series)

Holtz, S., *Corporate Conversations: A Guide to Crafting Effective and Appropriate Internal Communications*. New York: AMACOM, 2004.

Jablin, F. M. and Putnam, L., *The New Handbook of Organizational Communication: Advances in Theory, Research, and Methods*. Thousand Oaks, CA: Sage Publications, 2001.

Levine, R. V., *The Power of Persuasion: How We're Bought and Sold*. Hoboken, NJ: John Wiley & Sons, 2003.

McLeary, J. W., *By the Numbers: Using Facts and Figures to Get Your Projects and Plans Approved*. New York: American Management Association, 2000.

Munter, M., *Guide to Managerial Communication: Effective Business Writing and Speaking*. 4th ed. Upper Saddle River, NJ: Prentice Hall, 1997.

New, C. C., and Quick, J. A., *How to Write a Grant Proposal*. Hoboken, NJ: John Wiley, 2003.

Pan, Y., Scollon, S. B. K., and Scollon, R., *Professional Communication in International Settings*. Malden, MA: Blackwell Publishers, 2002.

Pearce, Terry. *Leading out Loud: Inspiring Change through Authentic Communication*. New and rev. ed. San Francisco: Jossey-Bass Publishers, 2003.

Rankin, E., *The Work of Writing: Insights and Strategies for Academics and Professionals*. San Francisco: Jossey-Bass Publishers, 2001.

Ryan, K., *Write up the Corporate Ladder: Successful Writers Reveal the Techniques That Help You Write with Ease and Get Ahead*. New York: American Management Association, 2003.

Simmons, J., *We, Me, Them, & It: The Powers of Words in Business*. New York and London: Texere, 2002.

Stockard, O., *The Write Approach: Techniques for Effective Business Writing*. San Diego: Academic Press, 1999.

Tingley, J C., *The Power of Indirect Influence*. New York: AMACOM, 2001.

Wang, J. X. and Roush, M. L., *What Every Engineer Should Know about Risk Engineering and Management*, Boca Raton, FL: CRC Press, 2000.

Whalen, D. J., *I See What You Mean: Persuasive Business Communication*. Thousand Oaks, CA: Sage Publications, 1996.

Wiener, V., *Power Communications: Positioning Yourself for High Visibility*. New York: New York University Press, 1994.

Weissman, J., *Presenting to Win: Persuade Your Audience Every Time*. Upper Saddle River, NJ: Financial Times/Prentice Hall, 2003.

Worth, R., *Webster's New World Business Writing Handbook*. Indianapolis: Wiley Publishing, 2002.

Index